高等院校土建学科双语教材（中英文对照）
◆ 土木工程专业 ◆
BASICS

木结构施工
TIMBER CONSTRUCTION

[德] 路德维希·史泰格　编著
李吉涛　李　鹏　译

中国建筑工业出版社

著作权合同登记图字：01-2007-3335号

图书在版编目（CIP）数据

木结构施工／（德）史泰格编著；李吉涛，李鹏译. —北京：中国建筑工业出版社，2011

高等院校土建学科双语教材（中英文对照）◆ 土木工程专业 ◆
ISBN 978-7-112-11875-5

Ⅰ.木… Ⅱ.①史…②李…③李… Ⅲ.木结构-工程施工-高等学校-教材-汉-英 Ⅳ.TU759

中国版本图书馆CIP数据核字（2010）第037571号

Basics：Timber Construction ／ Ludwig Steiger
Copyright © 2007 Birkhäuser Verlag AG（Verlag für Architektur），P. O. Box 133，4010 Basel，Switzerland
Chinese Translation Copyright © 2011 China Architecture & Building Press
All rights reserved.
本书经Birkhäuser Verlag AG出版社授权我社翻译出版

责任编辑：孙　炼
责任设计：姜小莲
责任校对：兰曼利

高等院校土建学科双语教材（中英文对照）
◆ 土木工程专业 ◆

木结构施工

［德］路德维希·史泰格　编著
李吉涛　李　鹏　译

*

中国建筑工业出版社出版、发行（北京西郊百万庄）
各地新华书店、建筑书店经销
北京嘉泰利德公司制版
北京云浩印刷有限责任公司印刷

*

开本：880×1230毫米　1/32　印张：5⅜　字数：156千字
2011年5月第一版　2011年5月第一次印刷
定价：**18.00**元
ISBN 978-7-112-11875-5
　　　（20279）

版权所有　翻印必究
如有印装质量问题，可寄本社退换
（邮政编码100037）

中文部分目录

\\ 序　7

\\ 导言　97

\\ 建筑材料　98
　　\\ 木材　98
　　　　\\ 木材的成长　98
　　　　\\ 木材的潮湿　100
　　　　\\ 切割类型　101
　　　　\\ 特性　102
　　　　\\ 承载力　103
　　\\ 木结构产品　104
　　　　\\ 实木　105
　　　　\\ 木产品　107
　　　　\\ 结构木板　110
　　\\ 木材防护　111
　　　　\\ 精选木材　111
　　　　\\ 结构性防护　111
　　　　\\ 化学性防护　112

\\ 木结构　113
　　\\ 结构稳定性　113
　　　　\\ 承重结构　113
　　　　\\ 加固　113
　　\\ 木结构体系　114
　　　　\\ 原木结构　115
　　　　\\ 传统木结构　119
　　　　\\ 木框架结构　121
　　　　\\ 木骨架结构　126
　　　　\\ 木面板结构　129

\\ 构件　131
　　\\ 基础　131

\\ 板式基础　131
\\ 条形基础　132
\\ 独立基础　132
\\ 墙脚　133

\\ 外墙　135
\\ 层式结构　135
\\ 建造科学　136
\\ 外部覆层　138
\\ 表面处理　146
\\ 内部覆层和设备安装　146
\\ 孔洞　148

\\ 内墙　150
\\ 结构　150
\\ 安装固定　151

\\ 顶棚　153
\\ 托梁顶棚　153
\\ 结构　153
\\ 托梁　154
\\ 底板　156
\\ 实木顶棚　158

\\ 屋顶　160
\\ 坡屋顶　161
\\ 平屋顶　165

\\ 结语　167

\\ 附录　168
\\ 标准　168
\\ 参考文献　169
\\ 图片出处　169

CONTENTS

\\Foreword _9

\\Introduction _10

\\Building Material _11
    \\Wood _11
        \\Growth _11
        \\Timber moisture _13
        \\Types of cut _15
        \\Properties _16
        \\Loadbearing capacity _18
    \\Timber construction products _20
        \\Solid wood _20
        \\Timber-based products _22
        \\Structural board _26
    \\Timber protection _27
        \\Choice of wood _27
        \\Structural timber protection _28
        \\Chemical timber protection _28

\\Construction _31
    \\Structural stability _31
        \\Loadbearing system _31
        \\Reinforcement _31
    \\Timber construction systems _33
        \\Log construction _34
        \\Traditional timbered structures _38
        \\Timber frame construction _40
        \\Skeleton construction _46
        \\Timber panel construction _49

\\Components _52
    \\Foundations _52
        \\Slab foundations _52
        \\Strip foundations _53
        \\Individual footings _54
        \\Base _54
    \\External wall _57
        \\Layered structure _57
        \\Building science _57

\\External cladding _60
\\Surface treatment _68
\\Internal cladding and service installation _70
\\Apertures _71
\\Internal wall _73
\\Structure _73
\\Fixing _74
\\Ceilings _76
\\Joisted ceilings _76
\\Structure _76
\\Joists _78
\\Seating _79
\\Solid ceilings _82
\\Roofs _85
\\Pitched roofs _85
\\Flat roofs _91

\\In conclusion _94

\\Appendix _95
\\Standards _95
\\Literature _96
\\Picture credits _96

序

  木材是人类最古老和最基本的建筑材料之一，直到如今，它仍然有很强的吸引力和实用性。在诸多文化氛围和气候条件下，建筑材料的选择，木材要优于砖材，在建材中具有重要地位。木材是一种生动的、轻质的、便于加工的材料，木材建造的房屋能够充分体现木材的这些特性。但是，木结构也有自己独特的区别于其他建筑材料的特性。所以，建筑师需要了解木材的特殊知识、木结构施工原则和规范，这样才能设计出合理使用木材的方案。

  在本套"国外高等院校土建学科基础教材"（中英文对照）系列丛书中，除了《砌体结构》和《屋顶结构》外，本书将继续给学生介绍木建筑研究的基本内容。在建筑课程中，学生们面对的第一个设计通常是木结构房屋，对于学习建筑方法和原则，木材是非常理想的材料，并且容易实践。因此，作者首先介绍木材的特性，它既是一种天然的建筑材料，也可以加工成建材产品；然后介绍木结构常见的施工体系和它的特殊规程，这种木结构规程应用于木结构所有构件连接和节点，并且通过举例来详细说明。

  本书能够让学生对木结构体系有大概了解，并深刻详细地区分了它们之间的不同之处。如果掌握了这些知识，就能够选择最合理的结构体系以满足我们的设计，并能够建设性地运用我们的建筑知识。

编者：**Bert Bielefeld**

FOREWORD

Wood is one of humankind's oldest and most elemental building materials, and has lost none of its appeal or validity. In many cultures and climates timber dominates over brick as the choice for house building. Wood is a living, light, simply worked material, and houses with a character all their own can be built from it. But timber construction has some particular characteristics that make it unlike other materials in construction. So architects need special knowledge about wood and the rules for timber construction, in order to develop quality designs that do justice to the material.

In addition to *Basics Masonry Construction* and *Basics Roof Construction*, the present volume in this series for students continues with the essentials for timber construction studies. The first designs in an architecture course are often for timber houses, as this material is ideal for learning construction methods and principles in a way that is close to practice. The author therefore begins by explaining the qualities of timber as a natural building material and the construction products developed from it, then moves on to the commonest timber construction systems and their specific rules. The construction rules learned in this way are then applied to all the connections and transitions for the building components, and are elaborated using examples.

The timber construction volume enables students to gain a general insight into individual timber construction systems, to understand them in detail and to distinguish them from each other. Armed with this knowledge, you can select the most sensible system for your design and apply your knowledge constructively.

Bert Bielefeld, Editor

INTRODUCTION

In a 1937 essay on training architects, Mies van der Rohe said: "Where does the structure of a house or building show with such clarity as in the timber structures of the ancients, where do we see the unity of material, construction and form so clearly? Here the wisdom of whole generations lies concealed. What a sense of material and what expressive power speaks from these buildings! What warmth they exude, and how beautiful they are! They sound like old songs." This statement by one of the 20th century's most important architects conveys both the fascination of timber construction and the challenge it presents.

The living material, the different kinds of timber, the large number of timber construction systems, the sophisticated stratification of the building components and the way they are jointed require a great deal of knowledge if this building material is to be used appropriately in student design work.

Unlike the monolithic massive construction procedures with which students are familiar, timber construction works by assembling members, following a fixed order, and working with a defined structural grid. In terms of planning, this means a more systematic approach is needed, and also a greater degree of detailing and drawing work. This book introduces students to timber construction in three stages. First, readers are familiarized with the material wood and its properties, then the most important construction systems and their characteristic joints, finally assembling components and fitting them together. Our presentation is based around simple, manageable building solutions that are suitable for identifying the key problems of any particular timber structure. Large-scale loadbearing systems, bridges or hall structures that are ideally suited to timber construction are not considered, but information on further reading is provided.

One particular difficulty in presenting timber construction should be mentioned, although it can also be seen as a great opportunity. Timber construction techniques can be said to be in a state of flux. To complement the existing traditional systems, the industry is introducing a large number of new materials and technologies to timber construction.

This book aims to structure this very broad field and provide an overview. This will involve first of all passing on established knowledge and tried-and-tested structures, but there will be at least an indication of new building materials and technical developments.

BUILDING MATERIAL

WOOD

Several hundred varieties of wood are used on a large scale all over the world. They all look different and have their own particular properties. Many of them are used for finishing, and in furniture manufacture. Relatively little coniferous timber is used in wooden buildings, so beginners do not have to be timber experts in order to build with timber. The important thing is to understand its anatomical structure, and to know about the fundamental physical properties of this material.

Growth

When using wood, it is important to be aware that a piece of timber, a beam or plank is part of a vegetable organism, a tree, and that its growth and quality are influenced by its surroundings. No one piece of wood is identical to another. Its properties depend in the first place on the kind of tree, and in the second on its position within the trunk.

Cells

The trunk consists of longitudinal tubiform cells, which are responsible for transporting nutrients as the tree grows. The cell walls enclosing the tubiform cavity are made up of cellulose and lignin (filler substance). The structure of the cell walls and the cell framework determine the strength of the wood. Unlike building materials such as non-reinforced concrete or masonry blocks, wood has a directional structure, corresponding to the path taken by nutrients from the trunk to the branches.

Cell growth takes place around the centre of the trunk, called the pith cavity, the oldest part of the trunk. It takes place in the form of annual growth phases, generally lasting from April to September in temperate zones, and creates annual rings.

> \\Hint:
> The loadbearing properties of a timber construction component are fundamentally determined by the loading across the direction of the fibre, the grain, or parallel to it. Plans must therefore contain information about the installation direction. In sections, the hatching makes it clear whether the timber is cut across or parallel to the grain.

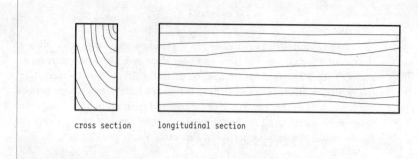

Fig.1:
Timber cut crossways and parallel

Early wood, late wood

Within these rings, the softer early wood is formed in the spring from large-pored cells, and the more solid late wood with thick-walled cells follows in the autumn. The proportion of late wood essentially determines the strength of the timber.

Sapwood, heart

This growth process can be read very simply by looking at the cross section of the trunk. According to the type of wood, the outer area, the <u>sapwood</u>, is more or less clearly distinct from the older, inner section, the <u>heartwood</u>. The heartwood has no supply role to fulfil, and is thus drier than the sap-bearing parts. Differences between heartwood and sapwood make it possible to divide timber types into:

- Heartwood trees
- Close-textured trees
- Sapwood trees

<u>Heartwood trees</u> have a dark core and light sapwood. They are considered to be particularly weather-resistant. They include oak, larch, pine and walnut. <u>Close-textured trees</u> show no colour difference between sapwood and heartwood, simply differences in moisture content. Both are equally light-coloured; the heart is dry, the sapwood moist. This applies to spruce, fir, beech and maple, for example.

<u>Sapwood trees</u>, on the other hand, show no difference in either colour or moisture content. They include birch, alder and poplar.

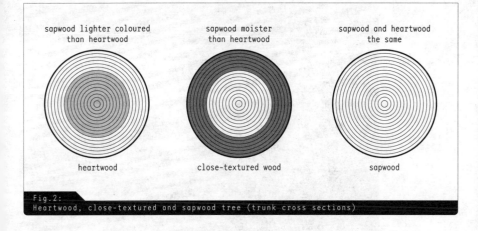

Fig.2:
Heartwood, close-textured and sapwood tree (trunk cross sections)

Timber moisture

Almost all physical properties of wood are influenced by moisture content. Its weight depends on this, its resistance to fire and pests, its load-bearing capacity and above all, its dimensional stability and consistency.

Shrinking, swelling

Wood swells and shrinks with changing moisture conditions. When wood dries, its volume is reduced, which is called shrinkage, and the reverse process, which causes an increase in volume, is known as swelling. This takes place because both the cell cavities and the cell walls contain water. As a hygroscopic material, wood is able to give off or absorb moisture according to the ambient conditions. This is also known as timber movement.

Moisture content must be specified for construction timber. Here a distinction is made between:

Green	more than 30% wood moisture
Semi-dry	more than 20% but maximum 30% wood moisture
Dry	up to 20% wood moisture

Construction timber should always be installed in a dry state, if possible at the moisture level expected at the location. Timber equilibrium moisture indicates the moisture levels at which only small changes of dimensions take place. For rooms this means:

Closed on all sides, heated	$9 \pm 3\%$
Closed on all sides, unheated	$12 \pm 3\%$
Covered, open	$15 \pm 3\%$
Structures exposed to weathering on all sides	$18 \pm 6\%$

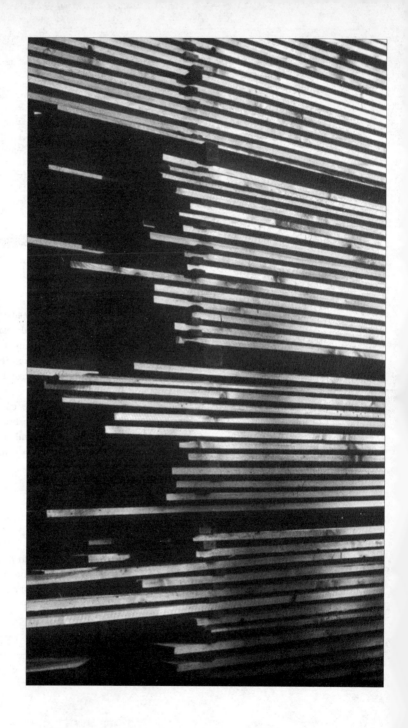

Wood moisture indicates the percentage of water contained, with reference to absolutely dry wood. But movement in wood is not a single, once-and-for-all process; it happens after the timber has been installed as well. According to the ambient atmospheric humidity, which is lower in winter than in summer, timber shrinks and swells seasonally as well.

Types of cut

Because of the difference between the water content of sapwood and heartwood, as well as between early and late wood inside the annual rings, shrinkage rates differ, and thus the cut timber becomes distorted. The key factor here is its position in the trunk.

Timber can be cut <u>tangentially</u> to the heart or <u>radially</u>, i.e. at right angles to the annual rings, and this affects the degree of volume change. According to the type of wood, the degree of shrinkage is usually more than double for tangentially cut than for radially cut wood. Longitudinal shrinkage is negligible.

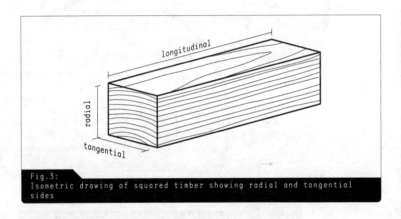

Fig.3:
Isometric drawing of squared timber showing radial and tangential sides

\\Hint:
One of the most important rules for timber construction is that wood must always be installed to allow for movement caused by shrinking and swelling, e.g. by leaving sufficiently large gaps between the timber components. Ideally a form board should be fastened with one screw only, in the middle or at the edge, so that the wood can move crosswise to the fibre direction (see chapter External walls).

The difference in volume change also means that planks or squared timber cut from one trunk at right angles distort differently. Tangential planks bend (bow) outwards on the side away from the heart, due to the shortening of the annual rings. Only the centre board, the heart board, remains straight, although it becomes thinner in the sapwood area. Figure 4 shows the reduction in volume (green colour) as cut timber shrinks.

Properties

Wood's finely porous structure makes it a relatively good material for insulation. The thermal conductivity coefficient of the coniferous timbers (softwoods) spruce, pine and fir is 0.13 W/mK, that of the deciduous timbers (hardwoods) beech and ash 0.23 W/mK. So in comparison with brick at

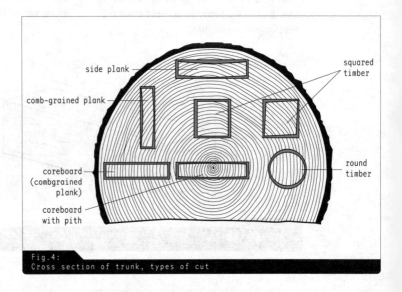

Fig.4:
Cross section of trunk, types of cut

\\Hint:
The side of a piece of tangentially cut timber furthest from the heart is designated the left-hand side, and the side facing the heart the right-hand side. The anticipated deformation should be taken into account when the timber is used for building.

0.44 W/mK or concrete at 1.8 W/mK, wood has considerably better <u>thermal insulation properties</u> than many other building materials.

In contrast, wood's <u>thermal expansion</u>, unlike steel or concrete, is so slight that it can be disregarded for building purposes.

Gross density

Wood has a low gross density, so its <u>thermal storage capacity</u> is less than for solid building materials such as masonry or concrete. The thermal storage coefficient of spruce and fir is 350 Wh/m³K, whereas that of standard concrete is 660 Wh/m³K. This is particularly problematic for summer thermal protection. The thermal compensation between cool night and daytime warming is less in timber than for solid structures. The lower gross density also means that wood has a low sound insulation coefficient, but absorbs sound well because of its open cells.

Thermal storage capacity and sound insulation can be achieved only by incorporating heavy building materials as well, i.e. materials with a greater gross density, such as plasterboards or fibre cement, in the walls, or by correspondingly heavy floor coverings.

Fire prevention

Although wood is a flammable material (normally flammable), its behaviour in fire is not as problematic as would seem at a first glance. Wood with a large cross section burns relatively slowly and evenly from outside to inside because of the accumulating charcoal layer, so that it takes time to lose its loadbearing capacity. This is quite different from a steel girder, for example, which is not combustible, but deforms at high temperatures and loses its loadbearing capacity.

The burning rate of wood becomes lower the moister the wood is. Across the grain the speed is around 0.6–0.8 mm/min for softwood, for oak about 0.4 mm/min. In addition, behaviour in fire depends on external form. The greater the surface area at the same volume, the lower the fire resistance. This is particularly marked in the case of shrinkage cracks in solid wood. For this reason laminated wood without cracks will resist fire for longer and times can be calculated more accurately than for solid wood.

\\Hint:
Gross density indicates the strength of a building material. It depends on the weight of the material, and is given in kg/m³. The gross density of softwood is 450–600 kg/m³, of European hardwoods 700 kg/m³, of overseas hardwoods up to 1000 kg/m³. In comparison with this, standard concrete comes in at 2000–2800 kg/m³.

So with appropriate dimensioning, wood can meet fire prevention requirements.

Loadbearing capacity

Unlike masonry, which is ideally suited for dealing with load and pressure, wood can absorb compressive and tensile forces to an equal extent. But because of the above-mentioned tubular cell structure, the direction in which the force is applied is crucial. Parallel to the grain, in other words along its longitudinal axis, wood can absorb approximately four times as much compressive force than across the grain. The response to tensile force is even more extreme. Figure 5 shows the appropriate strengths for coniferous wood (S 10) as <u>admissible tensions</u> in N/mm² according to German standards.

For construction this means that, as far as possible, the timber should be installed so that the load is placed on its efficient longitudinal axis, where it can absorb compressive and tensile forces.

In general, loadbearing capacity depends on the proportion of thick-walled timber cells, and thus on the density of the wood. Hard deciduous timber such as oak is thus particularly suitable for compressive loading, as a sill or threshold timber, for example, while long-fibred coniferous timber is more suitable for dealing with bending loads.

As a building material that grows and shows all the irregularities of nature, the expected loadbearing capacity of construction timber is not guaranteed from the outset. It is therefore sorted visually and mechanically according to certain characteristics such as the number and size of the branches, any deviations in the fibres, cracks, gross density and elasticity, and then graded for sale.

Sorting

In Germany, loadbearing construction timber is divided into three sorting classes or grades, and the strength to be used for static calculations is stated in terms of these grades. › Tab. 1

In other countries, standardization can be even more sophisticated. In America, all construction timber is marked with a stamp providing the fol-

---

\\Hint:
In Germany the fire resistance classes F30 B, F60 B or F90 B indicate that a component will retain its ability to function in case of fire for 30, 60 or 90 minutes, respectively.

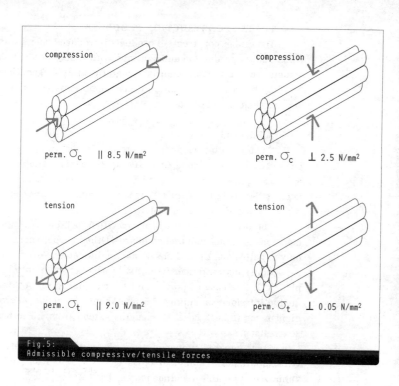

Fig.5:
Admissible compressive/tensile forces

Tab.1:
Sorting classes and grades in Germany

Sorting class	Grade	Loadbearing capacity
S 13	I	higher than average
S 10	II	normal
S 7	III	low

lowing information: grade, quality control organization, sawmill number, timber type, moisture content, E module, bending strength and use.

This makes allocating wood on the building site and building supervision control considerably easier.

TIMBER CONSTRUCTION PRODUCTS

The timber construction products introduced below start with solid wood and its production and continue with timber products in which the structure of the starting material is significantly changed. They finish with building boards, which contain other bonding materials and substances such as cement and plaster.

Solid wood

The term solid wood includes round timber with the bark removed or cut softwood or hardwood. Construction timber is available from sawmills as stock squared timber in particular cross sections and lengths. Timber is classified as laths, planks, boards and squared timber according to the ratio of thickness to width. > Tab. 2

Dimensions

The thicknesses and widths specified in Table 3 are the current dimensions for laths, planks, and boards. Other measurement systems differ only slightly from these dimensions in millimetres.

Construction timber is normally used as sawn, in other words unplaned. In the case of planks and boards planed on both sides, e.g. for visible interior use, approx. 2.5 mm should be deducted from the given dimensions in each case. The lengths extend from 1.5 to 6 m in 25 and 30 cm steps.

Stock squared timber is available in whole-centimetre dimensions with square and rectangular cross sections. The cross sections listed in Table 4 are the preferred dimensions.

American stock squared timber is based on the inch (25.4 mm) as a unit. Starting with a minimum width of 2 inches, slender cross sections emerge, principally adapted to American timber frame construction's close rib positioning. > chapter Timber frame construction and Tab. 5

The commonest timber types in Central and Northern Europe include spruce, fir, pine, larch and Douglas fir. In America they include Douglas fir, red cedar, Caroline pine and pitch pine.

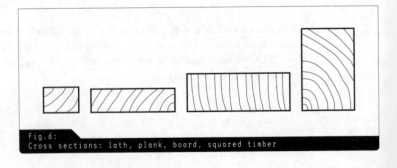

Fig. 6:
Cross sections: lath, plank, board, squared timber

Tab.2:
Cross sections for lath, plank, board, squared timber

	Thickness t / Height h [mm]	Width w [mm]
Lath	$t \leq 40$	$w < 80$
Plank	$t \leq 40$	$w \geq 80$
Board	$t > 40$	$w > 3d$
Squared timber	$w \leq h \leq 3w$	$w > 40$

Tab.3:
Customary timber cross sections

Lath cross sections	24/48, 30/50, 40/60
Thicknesses for planks	16, 18, 22, 24, 28, 38
Thicknesses for boards	44, 48, 50, 63, 70, 75
Thicknesses for planks/boards	80, 100, 115, 120, 125, 140, 150, 160, 175

Tab.4:
Customary dimensions of squared timber

6/6, 6/8, 6/12, 6/14, 6/16, 6/18
8/8, 8/10, 8/12, 8/16, 8/18
10/10, 10/12, 10/20, 10/22, 10/24
12/12, 12/14, 12/16, 12/20, 12/22
14/14, 14/16, 14/20
16/16, 16/18, 16/20
18/22, 18/24
20/20, 20/24, 20/26

Tab.5:
American timber sizes in inches

Widths	2, 2½, 3, 3½, 4, 4½
Heights	2, 3, 4, 5, 6, 8, 10, 12, 14, 16

Solid timber products

The following solid timber products are produced by further refining and finishing solid timber.

<u>Solid construction timber (SCT)</u> is sorted to the usual standards for strength, then also sorted and specially classified for appearance. It thus

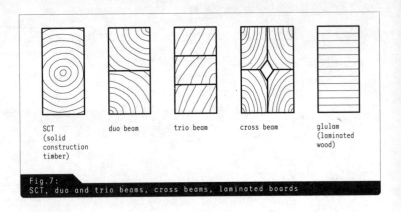

Fig.7:
SCT, duo and trio beams, cross beams, laminated boards

meets specific requirements in terms of loadbearing capacity, appearance, dimensional and formal stability, moisture content, limitations on crack width, and surface quality. Wedge finger jointing, i.e. gluing the wedge-shaped tenons on the ends, makes it possible to supply any required length. It is available in the customary cross section dimensions for stock squared timber.

Similarly improved solid timber quality is achieved with <u>duo</u> or <u>trio beams</u>, for which two or three boards or squared timbers are glued together on the flat side with parallel grain.

<u>Cross beams</u> are made up of quarter timbers glued together with parallel grain. Here the outside of the round timber segments is turned inwards, thus creating a central tube inside the rectangular cross section that runs through the full length of the beam.

<u>Glue-laminated timber (glulam)</u> meets very rigorous requirements in terms of formal stability and loadbearing capacity. It consists of softwood boards glued together under pressure, on the broad side, with parallel grain. Artificial resin adhesives based on phenol, resorcin, melamine or polyurethane are used to achieve waterproof adhesion. These show up in different colours externally in the glued joints, ranging from dark brown to light.

The boards are dried before gluing and planed, and any flaws in the wood are removed mechanically. The laminated gluing means that there is next to no deformation of the timber cross section. Glue-laminated timber is often used for wide loadbearing structure spans because it can be supplied in cross sections of up to 200 cm, and up to 50 m long.

Timber-based products

Timber-based products are a particularly economical use of wood, as they even redeploy timber processing waste such as shavings and fibres, as well as timber components such as boards, members, veneers and veneer strips.

Manufacture is industrial, by pressing with artificial resin adhesives or mineral bonding agents. This means that the original product is considerably enhanced. The irregular qualities of the wood are homogenized. The static properties and the tensions to which they can be submitted can be established much more precisely for timber products than for solid wood. Timber products also shrink and swell considerably less than solid wood.

Timber products are usually supplied in panel form in standard dimensions, e.g. panels 125 cm wide.

Gluing

Timber-based products are classified all over the world according to the way in which they are glued, to provide information about how the particular product responds to moisture. In Germany, the classification shown in Table 6 applies.

American timber construction has four gluing classes. > Tab. 7

Tab. 6:
Timber product classifications for Germany

V20	unsuitable for moisture
V100	suitable for short-term exposure to moisture
V100 G	suitable for long-term exposure to moisture; protected against fungi

Tab. 7:
Product classes in American timber production

Exterior	persistent exposure to moisture
Exposure 1	high resistance in periodic rain, not suitable for long-term exposure
Exposure 2	normal exposure to moisture
Interior	for protected interiors not exposed to moisture

\\Tip:
To guarantee economical exploitation of the panels, the construction grid must be fixed at the panel stage to allow as little waste as possible. Given a panel width of 125 cm, the same distance between axes, or half that at 62.5 cm or even one third at 41.6 cm will offer the most economical approach.

Both Exterior and Exposure 1 correspond to the German V100 classification, while Interior is comparable with V20.

Timber products can be classified by component nature as:

_ Plywood and laminated boards
_ Chip products
_ Fibre products

Plywood and laminated boards

Plywood and laminated boards consist of at least three layers of wood glued on top of each other, with the grain direction set crossways. This transverse arrangement of the layers, also known as <u>crossbanding</u>, prevents the wood from moving and gives the panel the necessary strength and stability in every direction. Plywood and laminated boards are thus particularly suitable for reinforcing timber structures and load-bearing walls. They can also be used for exterior work, provided the correct adhesive is used, although edges are particularly susceptible to damp, and should be covered or sealed if installed as façade panels.

For <u>veneer plywood</u>, the veneers are glued on top of each other in three, five, seven or nine layers, according to the thickness of the board (8–33 mm). Veneer plywood, with at least five layers and more than 12 mm thick, is also called multiplex board.

<u>Strip board and blockboard</u>, also known as coreboard, is plywood that consists of at least three layers with a middle layer of strips lying crossways to the covering veneers and giving the board particularly good loadbearing properties.

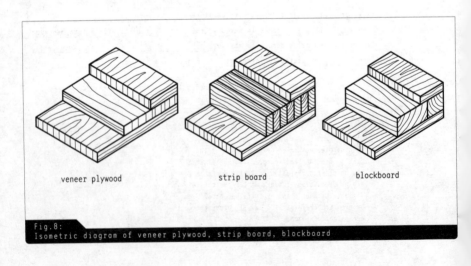

veneer plywood   strip board   blockboard

Fig.8:
Isometric diagram of veneer plywood, strip board, blockboard

The most commonly used plywood and laminated boards include:

_ Veneer plywood
_ Strip and blockboard
_ Three- and five-ply boards
_ Laminated boards
_ Multiplex boards

Chip products Chipboards use the waste products of the timber industry. Boards are made by <u>compressing sawdust and plane shavings</u> with adhesives. Unlike plywood, this does not produce continuous layers, but an intricate undirected structure. They can be used either as reinforcing planking › chapter Reinforcement for walls, floors, ceilings and roofs, or fitted in above-floor constructions as dry screed.

OSB A type of chipboard that is often used in the building industry is <u>oriented strand board</u>, so called because of its directed structure of relatively long (approx. 35 × 75 mm) rectangular chips or shavings. Because its direction changes layer by layer, it has directed mechanical properties like plywood. This means that very high strengths can be achieved, twice to three times as high as normal chipboard.

The following chip products are commonly used:

_ Flat pressboard
_ Laminated strand board (LSL)
_ Oriented strand board (OSB)
_ Extruded pressboard

Fibreboard The components of fibreboard are even smaller than for timber chip products. The original coniferous timber is so reduced in size that the wood structure is no longer recognizable. The product is manufactured <u>wet</u> without bonding agents using various procedures with and without pressure, and with and without adhesive, as follows:

_ Woodfibre insulation board
_ Softboard (SB)
_ Bituminized wood fibreboard
_ Medium hardwood fibreboard

or using a <u>dry</u> process, with adhesive added:

_ Medium density fibreboard (MDF)
_ High density fibreboard (HDF)
_ Hardwood fibreboard

Board made by the wet process is usually <u>softboard</u> for use in interiors, for sound and heat insulation and as roof formwork. <u>Medium density fibreboard (MDF)</u> is very popular for furniture construction and internal finishing because of its homogeneous structure. <u>High density</u> and <u>hard fibreboards</u> are used mainly for façade cladding.

Structural board

Board products that, unlike organically bound timber products, are bound inorganically are known as structural board. The starting material contains only as a certain proportion of wood, or none at all. They are divided into boards <u>bound by cement</u>:

- Cement-bound chipboard
- Fibre cement board

and boards <u>bound with plaster</u>:

- Plaster-bound chipboard
- Plasterboard
- Fibreboard

<u>Cement-bound boards</u> are characterized primarily by their high level of resistance to water and frost, and to attack by fungi and insects. They are therefore used largely as façade material, right down to the base area in contact with the ground. They are also suitable as reinforcing, effective boarding in timber construction.

On the other hand, <u>plaster-bound boards</u> are for interior use only, <u>plasterboard</u> mainly as cladding for walls and ceilings, and <u>fibreboard</u> commonly in several layers as screed in floor construction. Plaster- and fibreboard can be used for boarding external walls if they are permanently protected against the weather.

TIMBER PROTECTION

Pests

Unlike the mineral materials masonry and concrete, wood, as an organic material is subject to damage from vegetable (fungi) and animal (insect) pests. If it is attacked, the outward form of the timber can be affected, but also its loadbearing capacity, to the extent of completely destroying the building. Timber protection is thus of the utmost importance in timber construction.

Fungi need cellulose in order to develop. They flourish particularly in wet, warm, unventilated areas. Timber moisture content of at least 20% is needed for the ensuing wood rot.

Insects, mainly beetles, use timber in the sapwood area of conifers in particular to feed and house their larvae. Termites are among the insects that are most destructive of wood. They live mainly in the tropics and subtropics, as well as in America and the southern European Mediterranean countries. Infestation with termites is not visible from the outside: termites construct a system of passageways inside the wood in order to avoid loss of water. Buildings or furniture that have been infested collapse suddenly when a load is place on them.

Timber protection is known as control in the case of wood that has already been infested, or as preventive protection, to make sure that no infestation takes place.

When planning timber building, preventive timber protection is the most important. Essentially there are three measures available:

_ Choice of wood
_ Structural timber protection
_ Chemical timber protection

Choice of wood

Only timber that has been well dried and appropriately stored should be considered. Care should be taken that the moisture content is less than 20%.

Many countries' classifications also specify which types of wood are naturally resistant to insect damage. These include the heartwood ˃ chapter Growth of teak, greenheart, bongossi, oak and false acacia. In America, recommended types include the heartwood of black locust, black walnut and redwood. These woods can be used without chemical protection in parts of the building where particular exposure to moisture is expected. But chemical timber protection is essential if there is to be contact with soil.

Structural timber protection

Planners are the key figures in structural timber protection. The design, and particularly the details of it, should be laid out to avoid permanent moisture penetration of wood and wooden building sections. The topic will be discussed again in the detailed treatment of bases, windows and roof edges in the chapter Components.

Chemical timber protection

Chemical timber protection should be used only when all other timber protection methods have been exhausted.

Possible chemical timber preservatives include water-soluble, solvent-based or oily coatings or impregnation materials. To avoid environmental pollution, timber preservatives should be applied only in closed facilities, for example by boiler pressure processes or trough impregnation. Only cut surfaces and drilled holes may be treated on the building site.

Chemical timber preservatives are used according to the static function of the building components. Here we distinguish between:

_ Loadbearing and reinforcing timber sections
_ Non-loadbearing timbers that are not dimensionally stable
_ Non-loadbearing timbers that are dimensionally stable timbers for windows and doors

For <u>loadbearing members</u>, preventive treatment is necessary. Local laws determine whether this can be in chemical form, e.g. according to the degree of risk and corresponding allocation to a risk class.

No chemical preservation measures need be taken under certain conditions and when using resistant timber types.

\\Tip:
Not only in timber construction, but particularly there, the following points should be noted:
1. Keep damp out (roof projections, recesses, dripboards)
2. Ensure that water can run off (sloping horizontal surfaces)
3. Let air in to wood that has become damp (ventilation spaces)

\\Hint:
Timber components touching the soil are subject to the highest risk. Contact with the soil and thus permanent exposure of the timber to moisture should be avoided in any structure. A base zone of about 30 cm separating the wood from the ground is typical of timber structures.

Doors and windows

Windows and external doors are non-loadbearing but <u>dimensionally stable components</u> that permit very low tolerances if they are to function properly. They need particular protection against damp. But it is possible to manage without chemical treatment when heartwood with a certain permanent strength is used.

Non-loadbearing components that are <u>not dimensionally stable</u> do not have to work within close tolerances. These include overlapping timber cladding, fences and pergolas. They can be constructed without preservatives and without being painted, provided it is accepted that they will become increasingly grey. Chemical timber preservatives should under no circumstances be used on large areas of interior timber.

CONSTRUCTION

"Wooden buildings have to be constructed, stone buildings can be drawn," the Swiss architect Paul Artaria once wrote. This somewhat provocative statement was not intended to deny masonry construction its structural legitimacy, but rather to indicate the special requirements of building with wood, the logic of rod-based structures and the systematics of timber jointing.

STRUCTURAL STABILITY

Timber construction offers several answers or construction systems to meet the elementary requirement of structural stability, with specific loadbearing mechanisms and their associated nodal complexes at interfaces.

A brief introduction to the static requirements of timber construction systems is followed by an introduction to the most important timber construction systems in this second part of the book.

Loadbearing system

The structural stability of a building depends on various factors. First, the material used must have an adequate loadbearing capacity and appropriate dimensions to absorb the <u>vertical loads</u> from walls, roof and ceiling. The subsoil must also be able to take such loads.

<u>Horizontal forces</u> also affect all buildings, mainly from wind loads, but also impact loads that affect the structure horizontally.

Reinforcement

One way of handling horizontal forces is <u>restraint</u> or rigid fixing. ˃ Fig. 10 The supporting members are fixed into the foundations so that they are flexurally rigid, to prevent them from shifting sideways or becoming deformed. The simplest and most primitive form of restraint is ramming the sharpened tip of a timber support into the ground. In reinforced concrete construction, restraining the concrete columns in bucket foundations

---

\\Hint:
This subject is examined in detail in *Basics Loadbearing Systems* by Alfred Meistermann, Birkhäuser Publishers, Basel 2007.

has become accepted practice. Wood protection requirements make this kind of fixing problematical in timber construction.

The appropriate way to reinforce a timber construction is by making walls and ceilings into a rigid structure in all three dimensions. Imagine a cardboard box whose side walls can be pushed relatively easily into a rhomboid shape until the lid is added. It is only when this third section, the horizontal lid, is put in place that a stable, reinforced structure is created. › Fig. 9

> ◫
> Diagonal
> bracing

The basic element of a reinforced surface is the fixed triangle. In timber construction, it is relatively easy to fasten members together in the form of a triangle. This is known as triangular bracing, and makes the rectangular wall frame into a fixed plate.

Figure 10 shows the various ways of making a plate structure: a) using two compression members, whose competence alternates according to the direction of the force; b) using a compression/tension member, where the loading changes from compression to tension as the force itself changes direction; c) steel load cables can only be tension loaded. Here, too, the competence of the one or the other cable changes with the direction of the force.

A triangular effect can also be achieved with flat elements, diagonal boarding or by planking with timber-product boards admissible for reinforcement.

These procedures have marked timber construction design. Distinctions can be made between the timber constructions discussed below in terms of how their walls are reinforced. Diagonal bracing is most clearly visible in historic timber-frame buildings, but a crisscross pattern of steel cables is often part of the characteristic detail of modern skeleton construction.

> ◫
> \\Hint:
> A flat section of wall or ceiling is considered statically stable as a plate if it can absorb forces on its longitudinal axis without deformation. Resistance to the same transverse forces is considerably less, and transverse deflection results. The component is then loaded as a panel. The direction of the force applied decides whether a flat construction element functions as a plate or a panel (see Fig. 9).

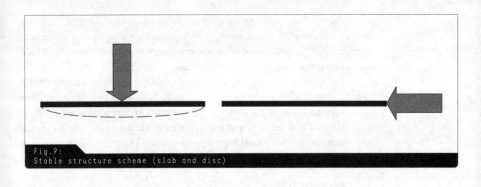

Fig.9:
Stable structure scheme (slab and disc)

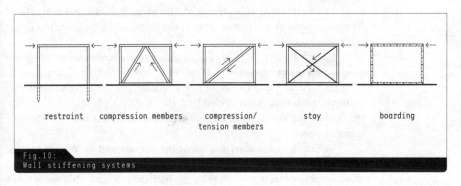

restraint   compression members   compression/tension members   stay   boarding

Fig.10:
Wall stiffening systems

TIMBER CONSTRUCTION SYSTEMS

Building methods develop out of different conditions: climate, cultural characteristics, the availability of material, tools, and the level of craftsmanship. Stone building developed in timber-poor southern Europe, while the wooded north produced timber construction. But here, too, there are regional differences. Log construction using solid timber walls developed in the mountainous regions of the Alps and the Mittelgebirge, as well as in northern Europe with its wealth of straight-trunked conifers. In contrast, the deciduous timber prevalent in central and eastern Europe led to traditional timbered building.

In the 19th century, and particularly the 20th, new technologies and materials changed European timber construction considerably. Construction engineering developed craft joints further to create high-quality steel jointing devices that made better use of the cross section of the timbers. These were used largely in skeleton construction. In North America, rib construction established itself as timber frame building using simple nailed joints.

The timber industry is constantly putting new natural and synthetic materials on the market. New transport methods and increasing demands for thermal insulation are also contributing to the evolution of timber construction.

To this day, the history of timber building remains a story of rod-based construction principles, from log to timbered construction, from timber frame to skeleton. Hence it is essential for everyone working in the field to know and understand the systems described below, even though new systems, mainly panel-based, will expand timber construction possibilities in future.

Log construction

The term "knitted construction" (Strickbau) is also found in specialist literature, because the beams cross at the ends, so can be described as being knitted together.

One characteristic of log construction is the large amount of timber needed and the great degree of slump of the horizontally placed members. Straight softwood that has grown very regularly is the most suitable. The walls were originally constructed from round trunks, slightly levelled off at the contact surfaces. The joints between the trunks were sealed with moss, hemp or wool.

The timbers are scarf-jointed at the corners and at the tying transverse walls, and generally anchored by cogging to make them tensionproof. The beams are offset to each other by half their height. This cogged link creates a kind of bond between the two walls. > Fig. 11

The degree of craftsmanship was also enhanced by the quality of the tools. Mortise and tenon improved jointing. Using squared rather than round timber evened out the cross section of the wall. Today's beams are elaborately profiled > Fig. 12

Joints

Typical timber construction joints are mortise and tenon between the horizontal trunks; scarf and cogging for binding walls at the corners of the building; and tensionproof dovetail joints for internal walls tying into the external ones. > Fig. 45, p. 74

A wall cross section consisting of single beams is no longer adequate for modern heating and cooling requirements. Modern log structures therefore have additional heat insulation. Ideally this is fitted on the outside, to avoid condensation. The horizontal timber look typical of log construction can then be achieved only by adding planks, which also protect thermal insulation material from weather. The timber industry now supplies laminated log construction wall in the form of sandwich elements of planks and insulation.

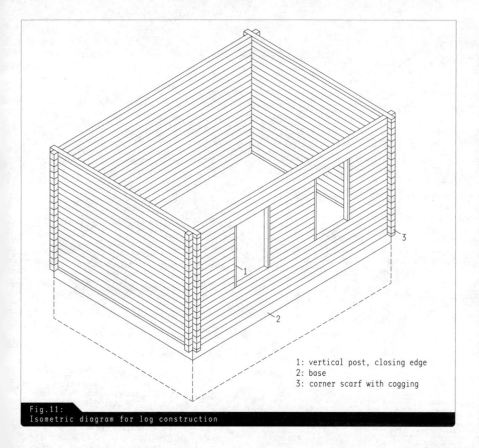

Fig.11:
Isometric diagram for log construction

1: vertical post, closing edge
2: base
3: corner scarf with cogging

Settlement

The compression load of the beams across the grain results in a high degree of <u>settlement</u> for log constructions, which can be up to 2–4 cm per floor. This factor must be taken into account when making windows and doors. The vertical framework or posts are therefore recessed sufficiently

> \\Hint:
> A scarf is a joint in which the timbers are recessed by half and fitted together flush. Tensile forces can be absorbed by additional grooving as a cogged joint. The same applies to the conical shape of a dovetail.

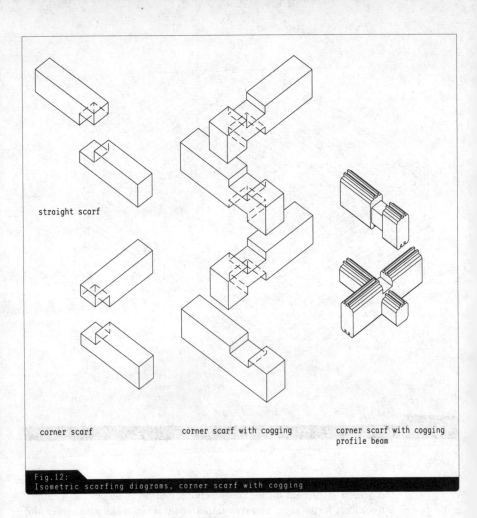

Fig.12:
Isometric scarfing diagrams, corner scarf with cogging

straight scarf

corner scarf   corner scarf with cogging   corner scarf with cogging profile beam

at the top to absorb the settlement of the wall without creating secondary bending. › Fig. 13 Settling should also be addressed by a concealed joint between window frame and lintel. For the same reason, any vertical runs through the building (chimney or services) should not be attached to the building rigidly, but in such a way that they can move. So although log construction many look primitive at first, it does require a great deal of craftsmanship and experience.

Log construction suggests a rigid, rectangular ground plan arrangement.

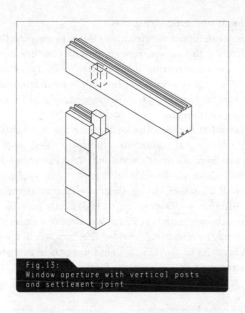

Fig. 13:
Window aperture with vertical posts and settlement joint

Fig. 14:
Log construction – external corner, internal wall connection, window aperture

The façade design should ensure that any apertures are as small and few as possible, so that the wall structure is not unduly compromised. A perforated façade of the kind familiar from classical masonry structures would be the appropriate design element for log construction, which is in essence a solid construction system. > Fig. 14

Traditional timbered structures

Traditional timbered construction clearly reveals the flow of forces in the structure. For this reason German specialist publications sometimes call it the "Stil der Konstruktion" (construction style). This kind of building is particularly attractive because of the visible distinction between loadbearing and non-loadbearing parts, between structural timbers and wall elements acting as filling.

Infilling

The spaces left between the loadbearing posts are called panels or compartments. Historical timber structures are filled with masonry, or with clay and wickerwork (wattle and daub). The current demand for warm interiors makes a heat insulation infill essential, with external cladding to protect it from the weather, but needing some internal covering as well.

As the filling is not loadbearing, it is perfectly possible to make apertures in the panels of a timbered wall. Windows cannot be placed at random, but they may be numerous provided they conform to the construction grid. Thus it is easier to provide daylight for rooms in timbered buildings than in log structures.

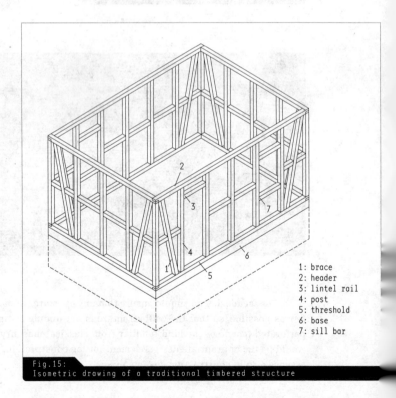

1: brace
2: header
3: lintel rail
4: post
5: threshold
6: base
7: sill bar

Fig.15:
Isometric drawing of a traditional timbered structure

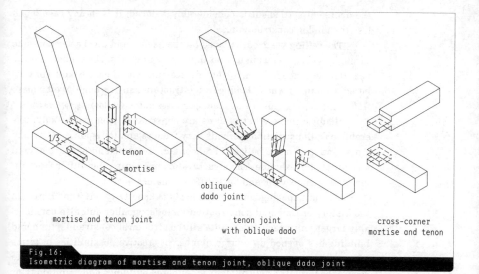

Fig.16:
Isometric diagram of mortise and tenon joint, oblique dado joint

Joints

> 🛈

Cross sections

Typical timbered construction joints are mortise and tenon, used to join timbers flush. An oblique dado joint is often used as well, to transfer forces better. Horizontal thresholds and headers are also joined by mortise and tenon or with a corner scarf. > Fig. 16 and chapter Log construction

One characteristic feature of timbered structures is that the vertical posts, horizontal rails and diagonal braces are held together at the bottom by the threshold and at the top by a header. As the timbers are mainly compression loaded, square cross sections with dimensions of 10/10, 12/12 or 14/14 cm are the norm. Horizontal rectangular cross sections are often used for thresholds and headers.

There are also timbered structures that run through one or more floors. As horizontal timbers are used only for the threshold and header,

> 🛈
> \\Hint:
> For a tenon joint the timbers are notched by division into thirds so that mortise and tenon fit together. The mortise should be less than 4 cm deep, so that the loadbearing timber cross sections are not weakened unduly.

Ceiling

the degree of settlement is considerably reduced; it is in any case much less than in log construction,.

The ceiling beams are placed on the header and can be seen outside through the beam ends in unclad timbered structures. The timbers for the next floor are then built on top of this, starting with the threshold. For timbered structures whose loadbearing structure runs through two or more floors, the ceiling must be suspended between the walls. › Fig. 50, p. 79

Historic timbered structures are constructed from hardwood, preferably oak. Construction methods vary from region to region. In Germany a distinction is made between Franconian, Alemanic and Saxon timbered structures, for example. The structural components often differ from region to region as well, and have different names.

Grid

The unit spacing between the posts is usually 100–120 cm. However, the history of timbered construction includes smaller and also considerably larger unit spacings. Despite all the structural constraints, timbered building has opened up a large number of creative possibilities in terms of both construction and design. Monument protection authorities try to secure the formal variety of historic timbered buildings and maintain their presence in the townscape. So when building within existing stock it is essential that architects be familiar with the principles of historic timbered construction. In modern timber frame building, the numerous posts, braces and rails raise the problem of filling a large number of joints with thermal insulation material, which creates a great deal of work. Nowadays the high level of manual craft input for timber jointing has largely been replaced by computer milling.

Timber frame construction

Modern timber frame construction originated in North America. Rapid settlement of the country along the new railroads demanded a simple, economical building method that could be carried out in a short time. Timber was the material that was available and suited to all the continent's different climatic conditions.

Industrial techniques started to influence timber construction in the first half of the 19th century. Steam-powered sawmills and machine-cut nails changed timber construction, which had until then been derived from traditional European timbered buildings.

Large numbers of different timber cross sections were replaced by uniform, plank-like cross sections. Simple nailed joints, easy to execute without special tools in an easy, do-it-yourself process, took over from elaborate manual jointing. The slender timber cross sections were nailed together laterally. The framework, more closely structured than in traditional timbered building, rose through the full height of the building. The

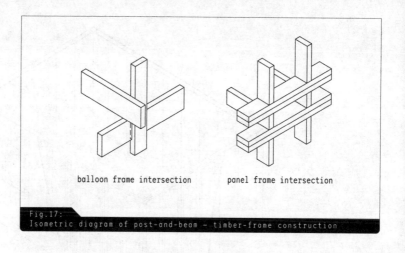

balloon frame intersection   panel frame intersection

Fig. 17:
Isometric diagram of post-and-beam – timber-frame construction

phrase "<u>balloon frame</u>" was first coined as a derisive term for the unusual lightness of the structure.

This building technique was known as <u>post-and-beam construction</u> in Europe. The disadvantages of post-and-beam construction, the difficulty of obtaining the timber, the additional difficulties caused by placing the high structural elements in position, and also problems with sound transmitted by the vertical members running through all the storeys, led to construction floor by floor, known as "<u>platform framing</u>" in America. The floor slab or ceiling then served as a working platform on which to assemble the framework.

<u>Timber frame building</u> developed from "balloon-frame" or post-and-beam construction. This is understood as a building method using wall elements assembled as frames while still lying on the ground. The frames are usually one storey high, although there are examples of two-storey timber frames.

Rib construction

All these construction techniques are sometimes collectively called <u>rib construction</u>, particularly in Germany (Rippenbau) because the upright members are so close together, and because of the narrow, plank-like cross sections.

Cross sections

Modern European architecture of the early 20th century was mainly focused on concrete as a material. Timber frame construction did not therefore become widely accepted until the 1980s, when American timber construction was remembered in the search for more reasonably priced building methods. The American standard "two by four" inches became 6 × 12 cm in Europe and, and thus produced somewhat sturdier timber

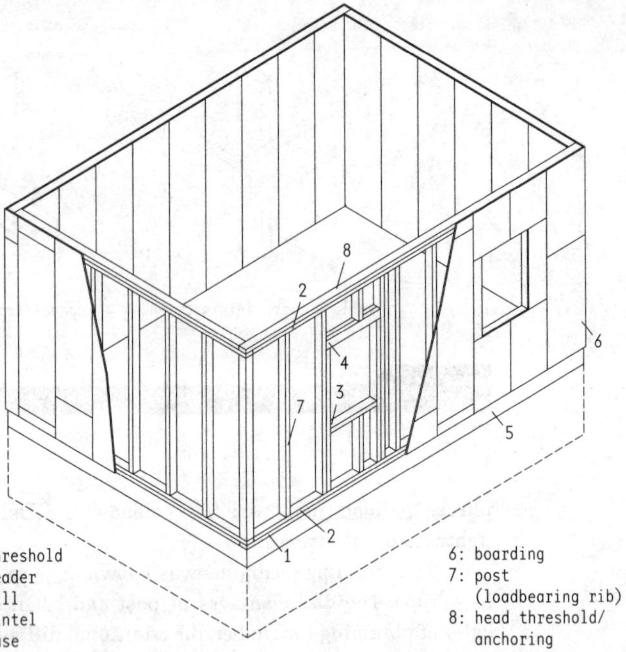
8
2
6
4
3
7
5
2
1
1: threshold
2: header
3: sill
4: lintel
5: base
6: boarding
7: post
(loadbearing rib)
8: head threshold/
anchoring
Fig.18:
Isometric diagram of timber frame construction

cross sections than the American dimensions, which work out at about 5×10 cm.

Joints

Typical wood jointing for the timber frame construction method is a nailed butted joint for the timbers. Diagonal nailing is intended to create

\\Hint:
"Two by four" (inches) is a tried-and-tested cross section for timber construction and can be used and combined in a variety of ways. Sometimes "two by four" is used as a generic term for timber frame construction as such. The higher insulation standard in the central European climate usually means that the cross section is thicker today, more like a side ratio of "two by six".

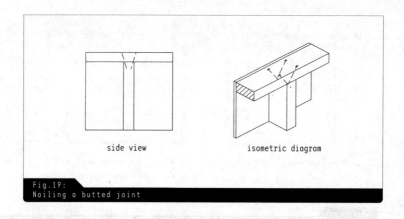

Fig. 19:
Nailing a butted joint

the maximum bond across the grain direction in the transverse joint. It is only the timber-product panel planking that makes the joint rigid and protects the nails from being pulled out. > Fig. 19

Grid

The tight rib spacing in timber frame construction with its small grid is usually matched to the width of the panels for the stiffening panelling. A popular unit spacing is 62.5 cm. > chapter Timber-based products Standard widths for insulation material could also be the deciding factor when dimensioning the construction grid.

One characteristic feature of timber frame construction is that the overall length of a building does not have to relate strictly to multiples of the unit spacing. The repeating pattern > Fig. 20 of the construction grid is often abandoned at the end of a wall and concluded with a special unit dimension. The arrangement of the windows and tailing internal walls can be handled freely as well; their positions are determined solely by the design and not by the construction grid, as would be the case for traditional timbered building. The construction grid in timber frame construction aims principally at using materials economically, rather than with structural and aesthetic order. Thus in comparison with other timber construction systems there are scarcely any restrictions on designing the ground plan or sections of the building.

Assembly

At the assembly stage, the structural timbers are no longer fitted together upright, as in traditional timbered building, but nailed together on the ground to form a <u>frame</u>, and subsequently set up on the threshold, which is firmly anchored to the floor slab, and fixed with nails. The double <u>threshold</u> made up of threshold and frame timbers is a typical feature of timber frame building, which uses lateral connections rather than complicated timber jointing, in other words prefers doubling or even tripling the cross sections where necessary. > Fig. 21

43

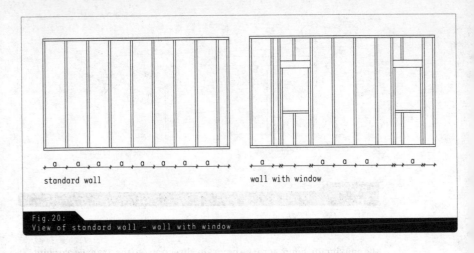

Fig.20:
View of standard wall – wall with window

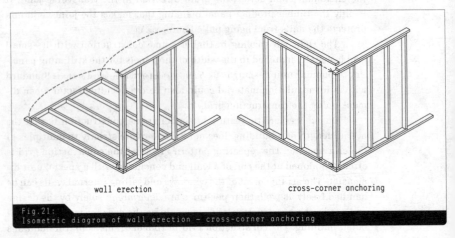

Fig.21:
Isometric diagram of wall erection – cross-corner anchoring

Walls that have been set up in this way are anchored on the same principle. The upper frames of the walls are fixed together using a second framing timber; the head threshold, which spans from one wall to the other, by a peripheral tie beam, which makes the structure tensionproof.

Ceiling

The course of beams then laid on the timber frame walls in the next step also consists of slender cross sections that correspond to the span of the ceiling. The beams are nailed to a set of beams of the same height running round the edge of the ceiling, and thus prevented from tilting. This joint is not fully stable until the ceiling is planked with timber-product panels. The timber beam ceiling thus forms a rigid plate, and can immedi-

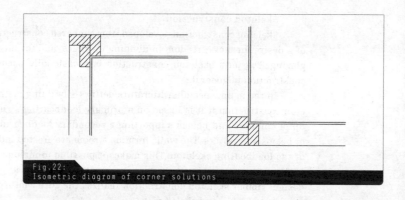

Fig.22:
Isometric diagram of corner solutions

Fig.23:
Timber frame wall – interior and exterior view

ately be used as a platform for the next floor, whose walls are set up using the same procedure. › Fig. 21 In America, buildings can be erected up to a height of six storeys by <u>platform framing</u>.

> ◊

◊
\\Hint:
When attaching walls, the corner posts must
be placed so that there is support available
in both directions for fixing the inner cladding for half the width of the cross section,
i.e. 3 cm. Possible solutions are shown in
Fig. 22.

Skeleton construction

Skeleton construction developed from timbered construction because of a desire for more freedom in dividing up space, and for larger areas of glazing. The term skeleton construction is occasionally used for timber construction in general.

In the main, specialist literature defines a building method as skeleton construction if it is based on a <u>primary loadbearing structure</u> made up of columns and beams supporting a secondary loadbearing structure of beams and rafters. The walls forming a room are erected independently of the loadbearing skeleton. This makes it possible to include large areas of façade glazing, as well as allowing greater flexibility in ground plan design. Timber skeleton construction realizes the 20th century Modernist principle of "<u>skin and skeleton</u>".

Grid

The spaces between the loadbearing posts are considerably larger than in timber frame construction, but also larger than in traditional tim-

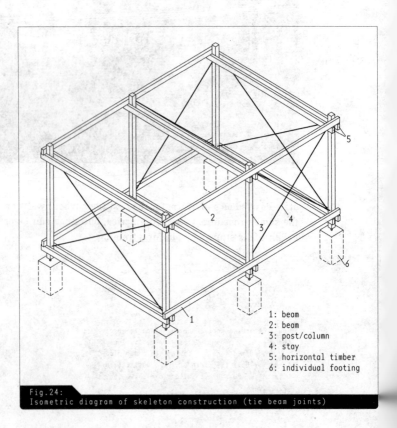

1: beam
2: beam
3: post/column
4: stay
5: horizontal timber
6: individual footing

Fig.24:
Isometric diagram of skeleton construction (tie beam joints)

bered structures. The loadbearing skeleton usually remains visible inside or outside. Glued laminated timber plays an important part in skeleton construction, as beams made of this make the large support spacings possible. › chapter Timber construction products

Wind forces are usually absorbed by steel cables or round steel bars arranged crosswise, as they can take only tensile forces › chapter Reinforcement

The wide column spacing suggests foundations based on individual footings. As the column is usually independent of the wall, the support base is not clad, so its galvanized steel joint with the footing is a striking architectural detail in the timber skeleton structure.

For buildings with more than one storey, erection is not floor by floor, but uses continuous posts. The horizontal beams are either attached in two parts as tie beams › Figs 24 and 25, or in one part with a butted joint. › Figs 26 and 27

Joints

In skeleton construction the columns and beams are jointed with metal devices without particularly weakening the timber cross sections. Unlike the craft joints used for log and traditional timbered construction, they are dimensioned by the structural engineers on the basis of technical construction tables. They are thus known as <u>engineered connections</u>.

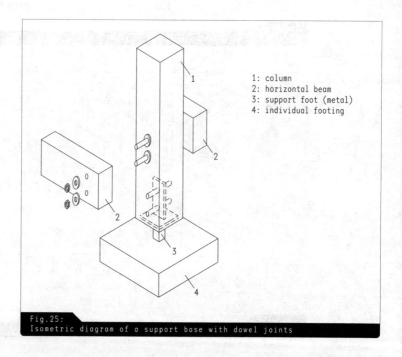

1: column
2: horizontal beam
3: support foot (metal)
4: individual footing

Fig.25:
Isometric diagram of a support base with dowel joints

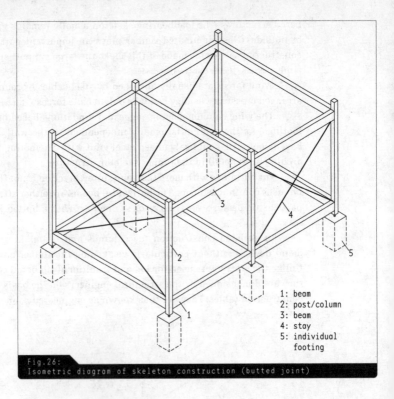

Fig.26:
Isometric diagram of skeleton construction (butted joint)

1: beam
2: post/column
3: beam
4: stay
5: individual footing

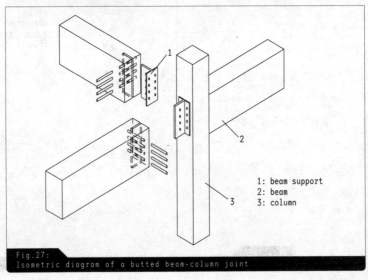

Fig.27:
Isometric diagram of a butted beam-column joint

1: beam support
2: beam
3: column

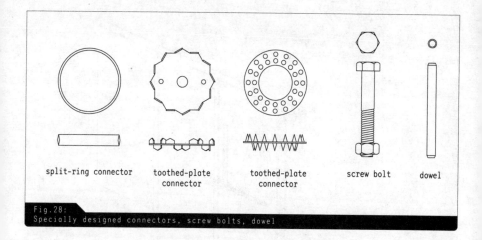

Fig. 28:
Specially designed connectors, screw bolts, dowel

Jointing materials

So that forces can be transmitted more effectively when timber components are joined laterally, › Fig. 25, <u>specially designed connectors</u> › Fig. 28 are usually inserted in a ring shape or driven in; these take the loads from the forces acting on the joint and transfer them to the maximum number of timber fibres. The joints are held together with <u>screwed bolts</u>.

Another kind of connector is the round steel dowel, which is driven into pre-drilled holes and transfers the load in association with steel plates that are mortised into the timber. In butted joints this means a connection can be made between beam and column that is largely invisible from the outside. ›Fig. 27, Fig. 50 and chapter Ceilings

Timber panel construction

The particular advantages of timber construction include relatively short assembly times and the dry construction system, which means that the completed structure can be occupied immediately. This short building period is considerably reinforced by the extent to which individual components, wall elements or entire room cells are prefabricated in the workshop.

Prefabrication

Timber frame building with its frame-like walls is particularly suited to prefabrication in the carpenter's workshop. Efforts to shift the maximum number of production phases from the building site to the workshop make the building process independent of weather conditions.

Timber panel construction maximizes this approach. The prefabricated panel units, usually a full storey high, are insulated and fitted with all the layers of building components and the external and internal cladding, so that on site they have only to be erected and fixed together. One essential

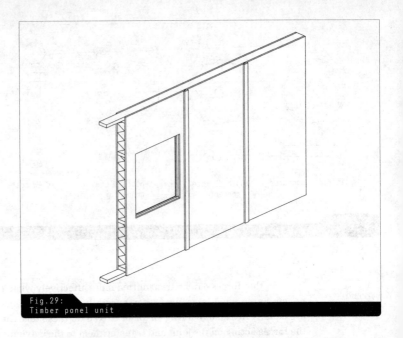

Fig.29:
Timber panel unit

detail is the jointing or unit butting. The floor and ceiling panels are also prefabricated and either placed on the wall units or suspended between them.

Timber panel building as a basic panel construction approach still requires a relatively large proportion of manual craftsmanship. But in recent years there has been a trend towards industrialization. Loadbearing wall units in solid cross-laminated plywood or edge-glued elements rather like oversized timber-product panels are produced by the timber industry for panel construction. It could almost be called slab construction.

This means that timber construction is increasingly turning away from stave to solid construction. For decades the rule was that the raw material wood should be used as sparingly as possible for timber construction. There now seems to be a change of direction, heralded by industrial building methods and a more consistent use of the raw material. › chapter Timber-based products

Slab construction

Slab construction using concrete panels is familiar from the Eastern European countries and Scandinavia. Because the concrete slabs are so heavy, unit size was restricted to a single storey. Timber is light in comparison, and permits units up to four storeys high. It is also possible to produce units to order, thus avoiding the danger of unduly stereotypical architecture, which is a general problem of prefabricated and unified façade units. Such production is made easier by CAD and computer-controlled machines.

Whether this slab construction method will become more generally accepted than stave construction for timber depends on various economic factors. Higher wage costs for stave construction have to be set against the disadvantage of elaborate and costly lifting devices, greater transport difficulties and large production shops.

COMPONENTS

This third section of the book addresses the essential points of detail for base, wall, ceiling and roof, and the way they are integrated into the overall timber construction pattern. Particular attention will be paid to the relatively complicated layer structure and to connections and junctions with adjacent building components.

Each of the building components is illustrated with detailed examples of solutions on a scale of 1:10 as timber frame or traditional timbered structures. These demonstrate contexts and problems, without claiming to come anywhere near covering the enormous range of possible detail.

FOUNDATIONS

As structural timber has to be protected, the foundations of any timber structure should raise it about 30 cm above the ground. The timber structure stands either on the ceiling of the cellar, or if there are no cellars, directly on a concrete or masonry foundation. Provided that the nature of the subsoil does not dictate special foundations, there are three basic foundation types available for timber construction: > Fig. 30

_ Slab foundations   = flat
_ Strip foundations   = linear
_ Individual footing = point

Slab foundations

Any timber construction system can use a foundation slab. Slab foundations are particularly appropriate for timber frame construction, which needs a platform to work on for assembly. The concrete floor slab can float either on a layer of compacted frost blanket gravel (coarse gravel) or on a continuous ice wall taken down to the frost line.

\\Hint:
This book will address roof construction and structures on roofs only to the extent necessary for external wall junctions, as roofs are dealt with in the separate volume *Basics Roof Construction* by Tanja Brotrück, Birkhäuser Publishers, Basel 2007.

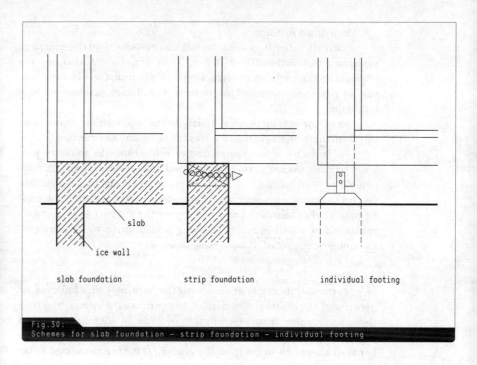

Fig. 30:
Schemes for slab foundation – strip foundation – individual footing

If there are cellars under the building, the cellar ceiling, usually made of reinforced concrete, replaces the foundation slab.

Strip foundations

If strip foundations are used a special floor structure must be made of timber planks. The crucial factor here is structural timber protection, which requires ventilation under the floor structure. The space between the ground and the wooden joists must be closed so that it is not accessible to small animals, but adequate cross-ventilation must also be ensured to keep air flowing around the timber joists.

Another important point is that thermal insulation is required for this type of foundation. After the timbers have been laid on the strip foundations the heat insulation can be inserted between the joists from above only. First, boards must be fixed on battens, which have previously been attached to the sides of the joists. > Fig. 33 One alternative to this building method would be to lay prefabricated insulated ceiling units.

In both cases the joists are dimensioned similarly to a normal floor unit, according to the size of the open span over the strip foundations

> chapter Ceilings

Individual footings

Individual footings are a foundation type particularly suited to skeleton construction. The building loads are concentrated on very few columns in skeleton construction, and are transferred to the subsoil at particular points. This reduces foundation excavations to a minimum.

The floor is structured similarly to the intermediate floors, usually with main and subsidiary joists for skeleton construction.

For reasons of structural timber protection, the industry provides various shaped parts in galvanized steel for the transition from column to foundations, > Fig. 31 the point at which the wooden column meets the concrete foundations. Despite this, particular attention must be paid at this sensitive point to keeping the support base as freely ventilated as possible and to allowing precipitation water to run off the timber unimpeded. > chapter Timber protection

Base

Particular demands are made on the base zone of a building regardless of the construction method. They are caused by moisture from the ground, spray from precipitation or snow in winter. For reasons of <u>structural timber protection,</u> > chapter Timber protection the external timber wall should be 30 cm from the ground. It is then connected to the ground with damp-resistant materials.

Constructing the base from exposed concrete is a common solution. The concrete provides protection from damp while giving the building a visible conclusion. For this reason the concrete foundation

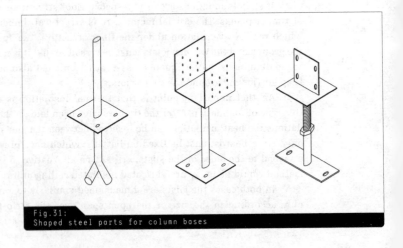

Fig. 31:
Shaped steel parts for column bases

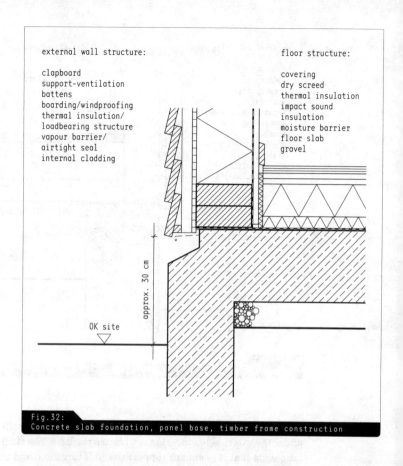

Fig. 32:
Concrete slab foundation, panel base, timber frame construction

slab is continued outwards to the façade plane. Using poured concrete, the base can be shaped so as to provide unimpeded ventilation behind the façade. To avoid ghost markings on the base surface, it should not be quite flush with the timber cladding, but very slightly recessed by the dimensions of a drip edging. The ventilated area behind the timber cladding must have an insect screen on the outside. › Fig. 32

Timber protection

Because it is so exposed and so close to the damp ground, a particularly resistant hardwood should be used as a threshold. Local legislation should be researched to determine whether it is possible to manage without chemical timber protection in this case, as would be permissible under German timber protection regulations. The timber must be separated from the damp concrete by an impervious course. The floor seal on the foundation slab and the vapour barrier membrane in the external wall are brought together at this point.

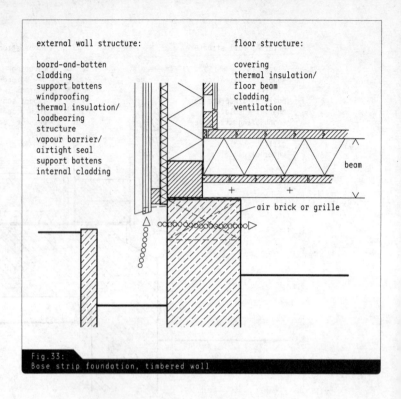

Fig. 33:
Base strip foundation, timbered wall

Figure 33 shows a solution in which the concrete strip foundation under the external loadbearing wall forms the base. The strip foundation is so wide that it is able to support the wall threshold and the joists under the floor structure. Then a surface humidity seal is not needed, but the cavity between the timber floor and the soil must be permanently ventilated. Ventilation grids or bricks should be incorporated into the foundations to ensure cross-ventilation under the building.

\\Hint:
Timber buildings must be anchored to the foundations. Both the threshold and the construction posts are attached to the floor slab or the strip foundations at regular intervals by connection anchors or heavy-duty dowels.

The visible base of the building is a key factor in determining the appearance of a timber structure. If, as in the case shown, the access height to the building is to be reduced, giving at least the impression of a building without a base, it is possible to lower the ground level in the immediate vicinity of the building. The base is set somewhat lower, in a trench running around the building, and there is still structural timber protection. For safety purposes it is best to cover this trench with a grille.

EXTERNAL WALL

As the building's envelope, the external wall has to withstand a number of stresses. It is affected from the outside by wind and rain, as well as by fluctuating temperatures, noise and radiation. From inside to outside the external wall has to contend with temperature gradient, air convection, sound transmission and water vapour diffusion.

Layered structure

In masonry construction, as well as in traditional log construction, one and the same material carries the load, insulates, seals and protects. The lightweight timber skeleton construction method distributes these tasks between various layers and specific materials. They have to be arranged in the correct sequence and matched to each other, as the system cannot accommodate omissions or weak points. The architect determines the structure, dimensions the thickness of the layers and clarifies the connections. The first decision to be made is on a single- or double-leaf structure. › Fig. 34

The fundamental difference between double- and single-leaf wall structures concerns the question of ventilation between the outer skin and the loadbearing wall. In a single-leaf structure the outer skin and the loadbearing wall together constitute a leaf; in a double-leaf structure the wall is divided into an inner and an outer leaf by the ventilation space. Each leaf has its own functions:

Outer leaf	Weatherproofing, ventilation
Inner leaf	Windproofing, thermal insulation/loadbearing structure, vapour barrier/air seal, airtightness, internal cladding

Building science

Ventilation

The safest, and thus also the most common version, is <u>double-leaf construction</u> with ventilation space. This space functions as an expansion chamber and provides pressure compensation for penetrating water. Care should be taken to ensure that precipitation water is able to drain away

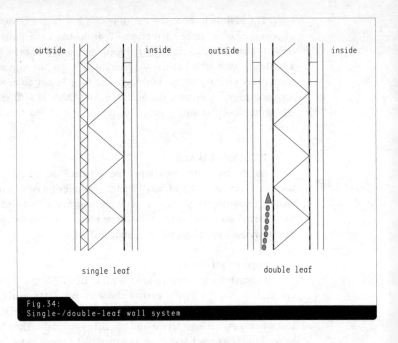

Fig.34:
Single-/double-leaf wall system

freely. At the same time the ventilation space means that water vapour from the interior of the building, or moisture from the thermal insulation, can be removed by the flow of air. Such ventilation is also advantageous in that it excludes summer heat by dispersing much of the warm air generated by the hot façade.

The ventilation space must be at least 20 mm deep, and must not be compromised by other items like windows and base. The air should come in at the base and be able to escape at the top of the wall, at the edge of the roof. The inlet and outlet apertures must be covered with an insect screen.

Windproofing

Windproofing material is applied to outside the thermal insulation. It prevents the thermal insulation from cooling off too quickly and stops air penetrating from the outside at the joints between the thermal insulation and the timber structure.

In the case of external cladding with open joints, › chapter External cladding the windproofing material must also protect the thermal insulation from damp penetration. To meet the requirement for increasing vapour permeability from the inside to the outside, the windproofing must however admit as much diffusion as possible. The timber-product stiffening panel takes over this function for structures whose loadbearing elements

Thermal insulation/loadbearing structure

have stiffening boarding on the outside if the butted joints have a rebate. Otherwise films or sheeting are used.

One of the reasons why timber construction is so popular with architects and their clients is that it copes considerably better than masonry with demands made on thermal insulation and increasing efforts to save energy. Timbers 12–16 cm thick with insulating infill provide good insulation values in their own right, although additional insulating layers are usually applied inside or outside.

Almost all the commercially available materials can be used for insulation. However, expanded and extruded polystyrene (EPS and XPS) and all <u>rigid foam panels</u> are problematic in that they cannot accommodate to timber shrinkage in the compartments. In this respect, <u>fibre insulating material</u> in the form of panels that can compensate for timber movement are better because they are easily compressed.

Insulation using <u>loose-fill cellulose</u> made of recycled paper is another common solution. It is blown in, and can be used only in closed cavities. This makes it particularly suitable for insulation in timber frame structures with closed chambers created by the posts and panelling.

Cellulose insulation absolutely must be protected from damp, as the walls could otherwise be deformed by the considerable increase in volume, causing structural damage that could be difficult to repair. Boron compounds are used to protect it against rot and high flammability.

Vapour-/airtightness

The chapter Timber protection showed in detail that timber must be protected against damp. This applies not only to damp penetrating from the outside, but also to water vapour inside the building, which accumulates in the structure as <u>condensation</u>.

Moisture can penetrate parts of the building by <u>diffusion</u> (vapour) or <u>convection</u> (interior air). Thus the need to be vapour- and airtight is a key

> \\Tip:
> The gaps between the loadbearing structure and the insulation can be bridged by additional internal or external insulation extending beyond the compartments. Low-density wood fibreboard panels (see chapter Timber-based products) are often used because of their relatively good inherent stability, and they are often used for windproofing and, in the bituminized version, for protection against damp behind the external cladding as well.

factor in timber construction. Vapour barriers and airtightness provisions are intended to prevent the insulating effect being lost by condensation or draughts as a result of permeability or leaks.

This applies to both single- and double-leaf systems. An airtight envelope is required, with as few penetrations or connections as possible. The vapour barrier and airtightness functions are generally combined in a single layer, which is placed inside the thermal insulation.

Thermal diffusion resistance

In principle, the external wall should be constituted so that vapour diffusion into the component is prevented and any water vapour that may already have penetrated is removed to the outside. Care should thus be taken when installed wall materials that the thermal diffusion resistance (S_d) decreases from the inside to the outside.

We distinguish the S_d values:

Diffusible	$S_d < 2$ m
Vapour retarder	S_d 2–1500 m
Vapour barrier	$S_d \geq 1500$ m

A <u>vapour retarder</u> is adequate for double-leaf construction. It is made of special paper or film and ensures that the water vapour generated within the building can diffuse to the outside, dosed and controlled by the thermal insulation, and is removed by the ventilation.

Single-leaf, unventilated structures need an internal <u>vapour barrier</u>. This is intended to prevent water vapour diffusing from the inside to the outside. The vapour barrier consists of vapourproof strips of plastic or metal film.

External cladding

The outer skin of timber or timber products provides weatherproofing for timber construction. But it is not uncommon to find cladding in other materials such as metal or plaster in some countries, such as America. For structural reasons alone it makes sense to combine materials with the same properties. This applies particularly to wood, which as a living material <u>swells</u> and <u>shrinks</u>. Design also thrives on the expressive quality of wood as a building material, its expressive properties, surface and texture.

Exterior cladding offers a large number of design possibilities. They include the choice of cladding type, the width of the boards or panels, their direction, the type of timber, surface treatment and patination.

Substructure

The substructure of loadbearing battens is invisible, but is still an essential component of the external cladding. The type of substructure depends on whether the cladding is ventilated, and whether it runs horizontally or vertically. It is attached to the loadbearing structure. The supporting battens are spaced according to the external cladding board thicknesses and vice versa. > Tab. 8

The principle of transverse arrangement requires horizontal <u>loadbearing battens</u> for vertical cladding. Here, vertical counter or <u>ventilation battens</u> are also required in order to guarantee uninterrupted air circulation from bottom to top, which horizontal battens would prevent. Vertical loadbearing battens fulfil this function for horizontal cladding.

Fixing

<u>Screws,</u> <u>nails</u> or <u>brackets</u> can be used to fix the boards. Nailing carries the risk of damage to the surface of the cladding and of the substructure. Screw fixing is safer and more readily controllable.

Corrosion-resistant materials are not necessary in every case, but nails or screws made of stainless steel or galvanized material are generally used to avoid rust marks on the surface of the wood. Fixing should be done so as to ensure that the wood is not prevented from shrinking and swelling. If cladding boards overlap, as in the case of board-and-batten or lap-joint cladding, the nails or screws have to past through one board only. Covering strips should be fixed to a single plank, or in the joint. The screwing or nailing should not pass through end-grain timber in secondary components either.

Tab. 8:
Batten spacings

Board thickness [mm]	Batten spacing [mm]
18	400–550
22	550–800
24	600–900
28	800–1050

\\Hint:
The S_d value of a component layer expresses its diffusion resistance as the thickness of a notional layer of air in repose with the same resistance. It is measured in metres and is the product of the thickness of the material (S) and its diffusion resistance μ. $S_d = \mu \times S(m)$. The greater the S_d value of a layer, the more vapourtight it is.

\\Hint:
The properties and effect of timber as façade cladding are studied in greater detail in the volume *Basics Materials* by Manfred Hegger, Hans Drexler and Martin Zeumer, Birkhäuser Publishers, Basel 2007.

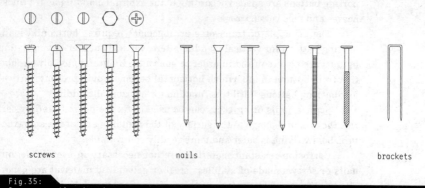

Fig.35:
Screws, nails, brackets

Vertical claddings include: › Fig. 37

_ Board-and-batten cladding
_ Coverstrip cladding
_ Lidded cladding
_ Matchboard cladding

Board-and-batten cladding

Floor and covering boards overlap by about 20 mm in board-and-batten cladding. Thus, when using boards of the same width, this produces a rhythmic visual impression of wide covering boards and narrower floorboards. One feature of this type is that, like coverstrip cladding, it has a relatively strongly profiled surface.

Coverstrip cladding

In coverstrip cladding, the gap of about 10 mm between the vertical cladding boards is covered by a strip in order to prevent penetration by precipitation water. › Figs 36 and 37

\\Tip:
In board-and-batten cladding the horizontal battening is sufficient as a support structure. There is no need for ventilation battens as the air cross section between the covering board and the inner board runs vertically and provides adequate ventilation for the outer skin (see Fig. 36).

\\Tip:
In board-and-batten cladding the boards should be placed with the heartwood side (see chapter Timber moisture) facing outwards, so that when the boards deform (bow) as they dry the joint between the floor and covering boards remains closed.

Fig.36:
Vertical cladding – lidded, coverstrip, strip and matchboard cladding

Lidded cladding

Matchboard cladding

The finishing strip is on the inside in lidded cladding, producing a relatively smooth, flush outer skin, as in matchboard cladding.

In matchboard cladding, the boards are joined by rebating or tongue-and-groove joints. This means that the boards can be fixed invisibly by concealed nails in the tongue or by using metal clamps. The boards must be left scope for movement by appropriate play in the connection between the boards. For matchboard cladding, additional boards or strips are needed to cover the open corner. In other forms of vertical cladding the corner is closed by a system of double layers of boards or strips. › Fig. 37

Horizontal cladding includes: › Fig. 38

_ Lap-joint cladding
_ Shiplap cladding

\\Hint:
Claddings in which the gap between the cladding units is closed are called closed cladding, contrasting with open cladding, which has open joints. For open cladding, windproofing must take the form of a moisture barrier to prevent precipitation water from penetrating the thermal insulation.

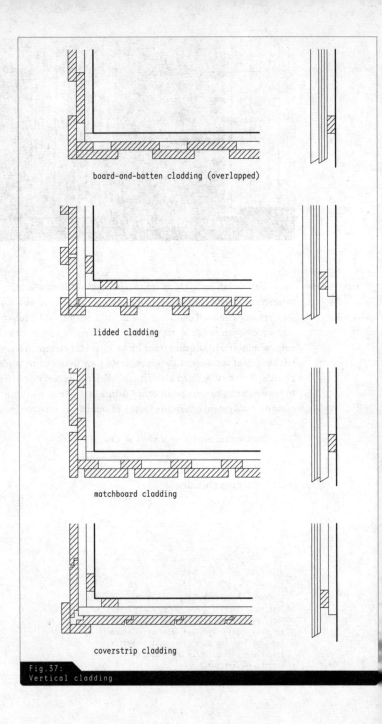

Fig.37:
Vertical cladding

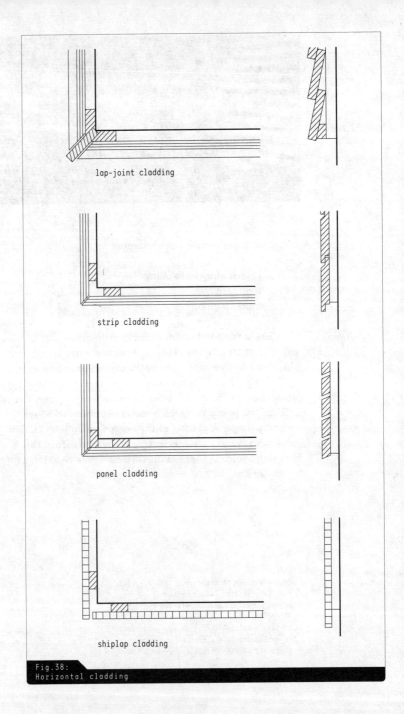

Fig.38:
Horizontal cladding

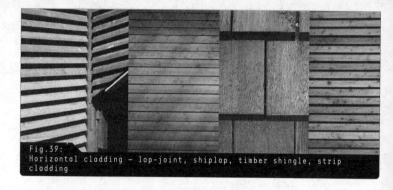

Fig.39:
Horizontal cladding – lap-joint, shiplap, timber shingle, strip cladding

And under certain circumstances:

_ Timber shingle cladding
_ Strip cladding
_ Panel cladding

Lap-joint cladding

The overlap in lap-joint cladding should be 12% of the board width, and at least 10 mm. The cladding boards are not usually rebated, but lap-joint cladding that is rebated in the overlap area does exist.

In lap-joint and board-and-batten cladding the overlapping principle provides scope for distributing the boards evenly over the surface of the wall and the points at which it meets doors and windows.

The diagonal placing of the boards in lap-joint cladding produces a geometrically difficult intersection at a mitred corner. This is often avoided by fitting a vertical board at the corner, to provide a finish for the cladding boards. > Fig. 38

\\Tip:
In the context of structural timber protection (see chapter Timber protection), horizontal cladding, unlike vertical cladding, offers the advantage that a damaged board at base level can easily be replaced. This is particularly important in the case of a shortfall in base height and exposed base areas.

Shiplap cladding

In contrast, a mitred corner is easily executed in shiplap cladding, and so is the concealed fixing. As in matchboard cladding, this can be carried out either with special metal clamps or by nailing or screwing in the tongue. > Figs 38 and 39

Timber shingle cladding

Timber shingle cladding is executed with small-format timber boards that are nailed or screwed like scales to supporting battens. > Fig. 39, centre right The intricate nature of this format (widths of 50–350 mm and lengths of 120–800 mm) means that it is easier to design curved forms or soft transitions. Timber shingles are available commercially sawn or split. Split shingles last longer, as the cell structure is not damaged by the splitting process. Gaps of 1–5 mm are left between the shingles to allow for swelling. They are often laid in two or three layers. Larch is an excellent timber for making timber shingles. If the roof is pitched steeply enough (30–40°), timber shingles can also be used for roof cladding.

Strip cladding

Strip cladding is a form of open cladding because the gaps are not covered. > Figs 38 and 39 The ventilation space is particularly important here for draining penetrating precipitation water away. In this case, the windproofing protecting the thermal insulation must also act as a moisture barrier. It makes sense to bevel the strips for structural timber protection. Strip cladding can also run vertically; bevelling the strips is then unnecessary.

Panel cladding

When using panel cladding, care must be taken to choose materials of the correct gluing class for the weather conditions. > chapter Timber-based products Timber-based products made of the following materials are suitable for external cladding:

_ Veneered plywood
_ Three-ply sheets of coniferous timber
_ Cement-bound chipboard

The edges of the timber products present a particular problem for panel cladding. One reliable solution is to cover the joints with a cover strip. This protects the sensitive panel edges. Some architects emphasize the structure of the cover strips in the façade by using strips of a different colour from the panels.

The flat, smooth effect made by panel cladding is most effective if the panel joints are executed as dummy joints. In this case the panels must be at least 10 mm apart. The panel edges must be protected from the effects of damp by water-resistant paint.

For horizontal joints, the underside of the panel should be undercut at an angle of inclination of 15°, so that any water running down can drip off. It must also be ensured that no water can stand on the upper edges of the panels. In addition, these edges must be covered, ideally with a metal element that is correspondingly angled outwards at 15°. > Fig. 40

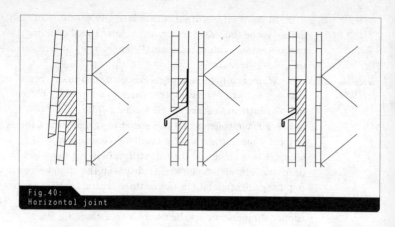

Fig.40:
Horizontal joint

The panels are usually fixed with visible stainless steel screws. A special substructure for hanging the panels is needed for invisible fixing. Systems are available commercially for this.

If the corners consist of a butted joint, as shown in Figure 38, care should be taken that the open edge does not face the prevailing weather.

Surface treatment

As external cladding is neither loadbearing nor dimensionally stable it does not need treatment with chemical timber preservatives. > chapter Timber protection

But physical timber protection is an option. This is intended to prevent intolerable quantities of precipitation water from being absorbed by the timber and to prevent the surface from UV radiation. Physical timber protection with pigmented transparent coatings or opaque paints prevents the timber surface from greying with time.

Greying

Timber greying is only a colour change: the timber is not actually damaged. Some architects deliberately use timber greying as a design

\\Hint:
As panel cladding is highly sensitive, preventive timber protection measures are advisable to reduce the impact of the weather, through the way the building faces, large roof overhangs or continuous balcony panels.

device. For example, the timber surfaces of the <u>Sea Ranch</u> by architects MLTW (Moore, Lyndon, Turnbull & Whitaker) on the west coast of America › photograph, p. 67 have acquired a silver-grey patina over the decades from UV rays and the weathering impact of the ocean, both of which have helped to make these buildings appear at one with nature. When planning detail, care should be taken that protrusions and recesses in the façade do not produce different weathering, and thus possibly undesirable colour effects.

Colour coating

An alternative here is to treat the timber with a coloured coating. Scandinavian timber buildings with strongly coloured paintwork show the impact such treatment can make.

<u>Pigmented transparent coatings</u> and <u>opaque paints</u> are both suitable for colouring timber. As a rule, painting schemes include undercoat, intermediate coat and topcoat. The crucial factor here is that the products used must be compatible with each other.

›

Internal cladding and service installation

Board cladding is one possible option for internal cladding, but timber products such as plywood can also be used. The sound insulation properties of the external wall are improved by one or two layers of plasterboard covering, also called <u>dry-wall finish</u>, because of their comparatively high gross density.

The internal cladding is applied either directly to the loadbearing structure or, preferably, fixed with battening underneath, thus making it simpler to position vertically, as well as enabling ventilation in the case of timber cladding.

Services layer

Installation work on the outside wall generally poses a problem for timber construction. To avoid penetrating the skin, which is closed in terms of building science (vapour- and airtight), all plumbing, heating and electrical services should be kept away from the outside wall and accommodated on inside walls wherever possible. But as it is almost impossible

> \\Tip:
> Cladding boards should be painted before assembly so that untreated areas do not appear when the timbers shrink. They should be painted with at least the undercoat on both sides to prevent the boards from bowing (see chapter Types of cut).

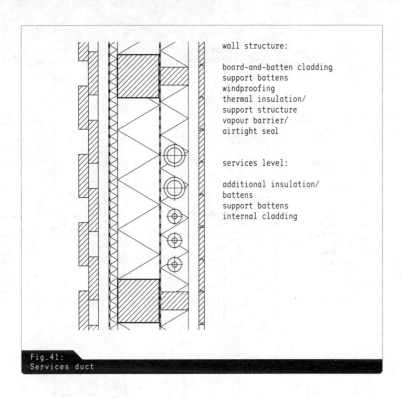

Fig.41:
Services duct

to avoid service runs in the external wall, they are housed in their own course in front of the vapour seal course. Space for pipework cross sections can be created by increasing the size of the batten structure to 4 or 6 cm. This internal shell can be used as an additional thermal insulation layer. Dedicated services walls and ducts should be planned for bathrooms and wet cells. › Fig. 41

Apertures

The wall's layered structure with windproofing, thermal insulation, vapour barriers and elements to ensure airtightness must be attached to apertures like doors and windows logically and carefully to avoid compromising the structural skin's effectiveness. The transition from the timber cladding deserves particular attention in terms of design as well.

The relatively highly profiled outer skin of lap-joint cladding is shown in detail in Figure 42 left, concluding in a frame running round all four sides, attached rigidly to the window frame. This board is placed diagonally at the lintel and the sill, so that rainwater can run down and drip off.

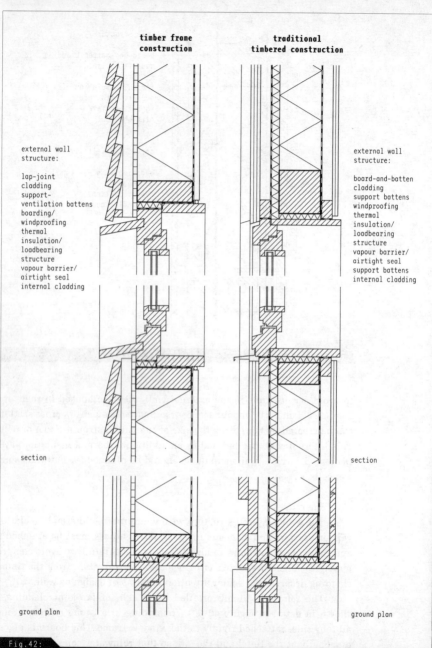

Fig.42:
Façade details window timber frame construction – traditional timbered construction

Ventilation   Ventilation for the timber cladding should also be ensured at the points where windows are attached. It must be possible for air to exit at the sill and enter at the lintel, and the ventilation space must be protected with an insect grille.

At the internal intersection, the gap between the window frame and the support structure must also be covered by a board running all the way round.

Vapour barrier   The vapour barrier must be attached directly to the window frame here, in the same way as the wind guard outside, and the gap between the window frame and the support structure filled with thermal insulation material.

The detail of a traditional timbered wall with board-and-batten cladding › Fig. 42 right shows that there is no need for a continuous external frame in this case, so the window has been shifted outwards onto the plane of the cladding. The vertical gap between cladding and window is covered with the external cladding board.

Wind sealing   The wind seal film is attached tight to the window frame using a batten. The frame must be rebated at the lintel so that the vertical cladding can be attached.

Inside, the deep reveal is contained by the window frame. The vapour barrier and the internal cladding are attached to this directly. The window positioned outside is attached to the loadbearing structure by galvanized steel brackets.

INTERNAL WALL
Structure

Loadbearing,   A basic distinction is made between loadbearing and non-loadbear-
non-loadbearing   ing internal walls.

Loadbearing internal walls carry their own load and that of the ceilings and roof. Reinforcing walls are also classed as loadbearing. They are part of the building's loadbearing system as a whole, and like the external wall must be constructed as a rigid wall plate, either by planking or braces. › chapter Loadbearing system The main function of non-loadbearing walls is to divide rooms.

As a rule, internal walls are constructed on the same <u>grid</u> and follow the same structural system as the external wall. The supporting ribs rise through the full height of a floor and are attached to the threshold and header. In a loadbearing wall, the header supports the ceiling joists. The space between the ceiling joists is finished like the internal wall structure.

Unlike the external wall, the main function of the internal walls is to protect against noise and fire. Thermal insulation is not a factor for walls dividing heated spaces inside the building, and their thickness does not have to take this into account.

Noise protection

The sound insulation value of a wall is determined in the first place by its weight per unit area. The sound insulation performance of an internal wall is raised according to the thickness of the planking with materials of the highest possible gross density, like plaster- or chipboard. <u>Cavity damping</u>, in which the panel cavities are filled with mineral or coconut fibre, makes a considerable contribution to sound insulation. It is sufficient to confine this to half to two thirds of the wall thickness, leaving room for <u>service runs</u>. The internal wall remains open on one side until the services are fully installed. If cellulose is being used for insulation, the insulating flakes cannot be blown in until the chamber is closed. If a particularly high level of sound insulation is needed, a <u>double-leaf</u> wall structure is required. Here, one leaf is articulated, preventing sound transmission from one side of the space to the other.

Fixing

Particular care must be taken when fixing the internal wall to the external wall. Both the external and the internal walls must be <u>anchored</u> non-positively, and an adequate basis for the internal cladding must be secured at the <u>inside corner</u>.

In <u>timber frame building</u> the walls are usually attached independently of the structural grid and determined solely by functional room division. Two additional posts are built into the external wall to enable a non-positive connection with the external wall, and also to create a means of fixing the internal wall cladding. › Fig. 22, p. 43 and Fig. 43

In <u>traditional timbered building</u> the distribution of the internal walls is matched to the <u>construction grid</u> for the loadbearing posts, so the wall junction lies on the axis of a timber frame post. This post is reinforced on both sides with a <u>wall stud</u> to which the internal cladding is attached.
› Fig. 44

\\Hint:
In damp spaces, specially glued timber-product panels or impregnated wallboard, identified by its green colouring in Germany, must be used. Gypsum fibreboard can be used without additional treatment in damp areas. Two layers of wallboard panels are required to support tiles if the ribs are more than 42 cm apart.

In log construction, the connection between the internal and external wall is made, similarly to the outer corner, with an <u>overlap joint</u>. The two walls are anchored so that they are tensionproof with a cog or a <u>dovetail joint</u>. ﹥ chapter Log construction The dovetail joint creates a characteristic feature for the internal wall because the end-grain is visible on the outside.
﹥ Fig. 45

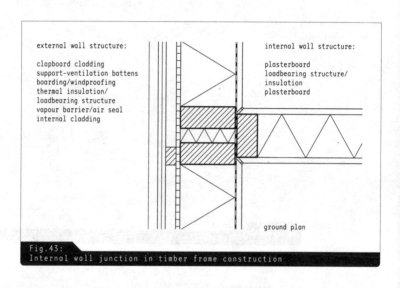

Fig. 43:
Internal wall junction in timber frame construction

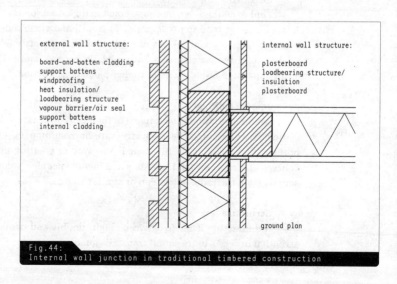

Fig. 44:
Internal wall junction in traditional timbered construction

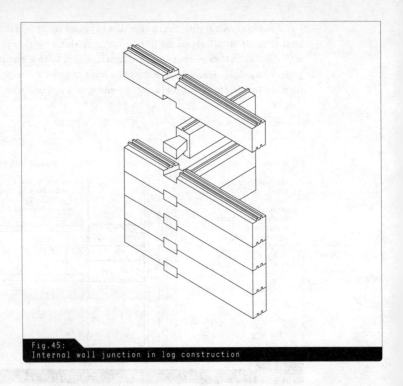

Fig.45:
Internal wall junction in log construction

CEILINGS

Timber ceilings can be constructed using either joists or solid wood. Joisted ceilings tend to predominate in traditional stave construction because of their economical use of wood. Ceiling systems made of prefabricated solid timber elements have recently become more common; they ensure speedier assembly, as in timber panel construction.

Joisted ceilings

Sound insulation

Sound insulation is an important factor in the construction of joisted ceilings. We distinguish between structure-borne or <u>impact sound</u>, and <u>airborne sound</u>. Walking on the floor is one way of creating impact sound. Sources of airborne sound transmission include people's voices in a room, sounds from the radio, television or similar sources.

Structure

A distinction is made between single, double and triple ceiling construction in terms of its sound transmission qualities.

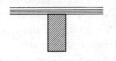
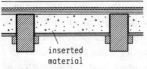
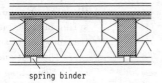

single

structure:
floor covering/
use surface
ceiling cladding
ceiling beams

double

structure:
floor covering/use surface
screed
impact sound insulation
ceiling cladding
insert (filling)
trickle protection film
ceiling joists
internal cladding (on battens)

triple

structure:
floor covering/use surface
screed
impact sound insulation
ceiling cladding
cavity insulation
floor beams
support battens (sprung)
internal cladding

Fig. 46:
Single, double and triple structures

Impact sound

In <u>single</u> construction there is direct contact between the floor covering and the support structure, and sound is thus transmitted freely when the floor is walked on.

In <u>double</u> construction the covering and support structure are separated by impact sound insulation. Here, the floor covering needs its own support structure, the screed, which in timber construction is in dry form. For example, dry screeds may be made from double-ply gypsum fibreboard or chipboard. › Fig. 46, Fig. 48 and chapter Timber-based products

\\Hint:
If the floor covering is separated throughout by the plate and the side walls, we speak of a floating structure. Here, it is important that sound is not transmitted via the walls even at the edges. A peripheral insulation strip should therefore be built in, as well as the horizontally laid impact sound insulation (see Fig. 49, p. 79).

Airborne sound

A triple structure is used to meet extra sound insulation needs, for example to seal bedrooms off from particularly noisy areas. Here, an articulated suspended ceiling is attached to the underside of the joists with spring fasteners (springy strips of sheet metal). This interrupts the direct transmission of sound waves generated by the vibration of the plate.

The first measure to be taken against sound transmission is to increase weight per unit area. Materials with a relatively high gross density are built into the ceiling, either on top of the ceiling cladding or inserted between the ceiling joists.

Insertion

Specially dried sand is a possible insertion material. › Fig. 46 It is inserted from above into cladding between the ceiling joists. This considerably reduces the visible height of the ceiling joists. A dividing sheet should be used to ensure that the sand does not trickle down through the joints when the ceiling vibrates.

Another way of increasing the weight of the ceiling on the ceiling cladding is to lay concrete paving stones, for example, and glue them firmly to the cladding. They are followed by the remaining floor structure, impact sound insulation, screed and the covering.

Joists

Spacing

When planning a ceiling with timber joists, care should be taken that the beams run parallel with the shorter sides of the room wherever possible. The maximum spacing for solid timber joints is about 5 m. Beams running over two or more fields are more economical than single-field joists.

Dimensioning

The loadbearing capacity of a timber joist is affected more by its height than its width. For this reason, timber joists are installed upright, so as to use the timber cross sections relatively economically. A side ratio of 1:2 or more is usual. The maximum height of squared timber is 240–280 mm.

TJI joists

TJI joists are a particularly efficient and reasonably priced system. They are used above all in American timber construction. The name derives from the manufacturer Truss Joint MacMillan Idaho. The double T structure is made up of solid or veneer plywood chords with a glued-in web of OSB panels. › chapter Timber-based products They offer very high loadbearing capacity combined with great lightness. As the members are very high, it is relatively easy to run services across the joists, and space is engineered for this purpose at the factory. TJI-joisted ceilings are usually clad on the underside. › Fig. 47

Space between joists

Joists are usually spaced at 60–70 cm intervals. But in timber construction it makes sense to match the joint spacing to the construction grid, so that where possible the ceiling load can be conducted directly into the loadbearing posts.

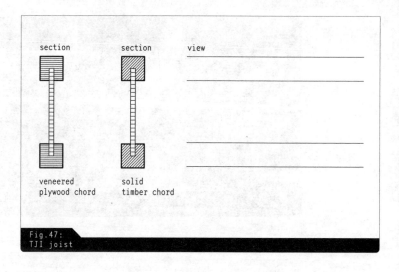

Fig. 47: TJI joist

In most floor-by-floor timber construction systems the walls support the joists. The necessary bearing depth can be calculated as joint height × 0.7.

> Trimmers

It is possible to use trimmers with loadbearing joists if they are penetrated by service shafts, chimneys or stairs, as is also the case of rafters in a timber roof truss. › Basics Roof Construction Trimmer joists are built in flush with the other joists, and attached to them by mortise and tenon. The joint is usually secured against being pulled out by a metal clamp. For fire prevention, there must be a clearance of 5 cm between the chimney and the timber joists.

Seating

As a detail typical of timber construction, › Fig. 49 the circular peripheral joist forms a kind of frame for the ceiling joists, securing the slender

> \\Hint:
> This formula gives a rough approximation for dimensioning timber-joisted ceilings:
> Joist height h = span/20
> Greater heights can be achieved only with laminated timber or timber products.

Fig.48:
Ceiling systems – joisted ceiling, joist skeleton construction, TJI joist

joists against buckling sideways while functioning as a compression and tension member similar to a peripheral tie beam in the system as a whole. To achieve a plate effect in the statical sense, the joists must be planked with panels that are suitable for stiffening, e.g. plywood. Care should be taken to offset the panel joints so that they can be bonded rather like masonry.

Vapour barrier

The continuous vapour barrier running through the external wall poses a particular problem. To prevent interruption of the barrier at the seating, the vapour barrier sheet should be taken round the edge of the ceiling and joined from one floor to the next. Care should also be taken that the layer of insulation is not interrupted or weakened at the seating, to prevent condensation water from accumulating either in the wall or at the seating. > chapter Building science

Joist hangers/ joist supports

> 🛠

The detail in Figure 50 shows the joists in a traditional timbered construction suspended between the walls. Steel joist hangers or joist supports are needed for this butted junction. These are available commercially in various sizes to suit the building statics.

The continuous rail at joist level makes a butted connection possible. In this case, the vapour barrier can run vertically on the same plane, but must be attached to the joist hangers or joist supports in such a way that the structure is vapourtight. In the detail shown a joist support is used because the timber ceiling joists are visible.

Internal seating

If the joists are not suspended as illustrated, but intended to lie on top of the walls, it is possible to create an internal seating in front of the external wall in the form of another internal supporting plane, which can

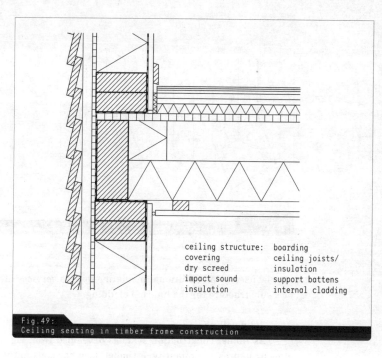

Fig.49:
Ceiling seating in timber frame construction

ceiling structure:
covering
dry screed
impact sound insulation
boarding
ceiling joists/ insulation
support battens
internal cladding

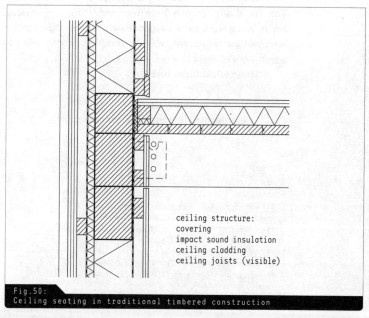

Fig.50:
Ceiling seating in traditional timbered construction

ceiling structure:
covering
impact sound insulation
ceiling cladding
ceiling joists (visible)

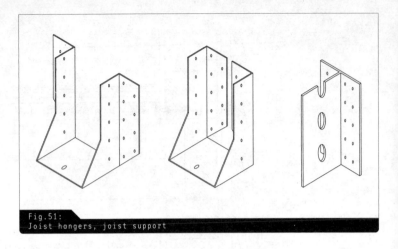

Fig. 51:
Joist hangers, joist support

also be used as an additional insulating layer, a services layer, › Fig. 41, p. 69 or a substructure for the internal cladding.

Solid ceilings

As timber construction systems developed from log to timber skeleton structures, the quantity of timber used has constantly been reduced. This trend has recently been reversed. › chapter Panel construction More timber is being used as the technology changes. This applies particularly to intermediate floors, for which a series of new systems have brought the advantages of solid floors.

These advantages include:

_ Shorter assembly times
_ Simple, usually industrial manufacture

> \\Hint:
> The use of visible joist hangers depends on whether the timber joist ceiling is clad on the underside. Otherwise, joist supports are used for visible joists. These fit into slots in the joist head and fasten the joist to the external wall using dowel pins placed transversely to the joint head (see Figs 28, and 48 centre).

- Thinner cross sections
- Increased thermal and sound insulation between floors

At the same time, the smooth lower edge of the ceiling makes the junction with the walls easier.

The additional structures on top do not differ in principle from those on joisted ceilings, although measures for improving thermal and sound insulation – i.e. inserts between the joists or surface structures to raise the unit area weight – are generally not needed.

Box beams

Solid wood constructions also include <u>box beams</u>, industrially prefabricated from boards and only assembled as a floor on site. They are supplied with a double mortise and tenon to this end. They are particularly suitable when large spans have to be bridged.

Industrial manufacture guarantees a precise fit and quality standard. The box beam units are a standard 195 m wide and are supplied in standard lengths up to 12 m. They are obtainable in heights from 120–280 mm according to span, graduated in 20 mm units.

The standard values in Table 9 apply to a load of 3 kN/m^2.

Tab. 9:
Standard values for dimensioning box beams

Span	Unit height
3.8 m	120 mm
4.5 m	140 mm
5.2 m	200 mm

Edge-glued construction

The edge-glued construction system uses the side boards from the trunk and cleared small dimension timber unsuitable for beams and squared timber. The boards are laid longitudinally and are joined upright without gluing by side nailing following a fixed nailing scheme to form ceiling or wall elements. Longitudinal joints in the boards have to be staggered. The elements, which are supplied with special rebates, are fastened to the complete section. › Fig. 52

Edge-glued elements must be protected from moisture, particularly during the building period, as water penetrating the boards would create a <u>degree of swelling</u> that is intolerable for any fitting situation.

They are supplied in standard dimensions according to manufacturer: in thicknesses of 100–220 mm, widths of up to 2500 mm and lengths of up to 17 m.

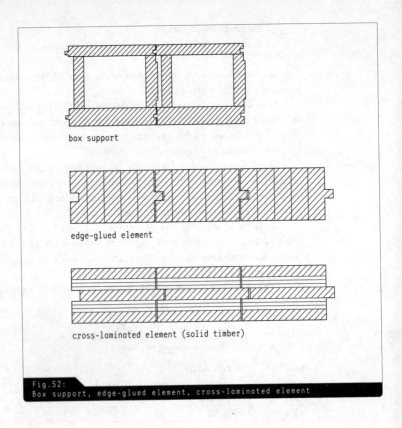

Fig. 52:
Box support, edge-glued element, cross-laminated element

Tab. 10:
Standard values for dimensioning edge-glued elements

Span	Unit height
3.6 m	100 mm
4.3 m	120 mm
5.0 m	140 mm

The standard values in Table 10 apply to loads of 3 kN/m².

Cross-laminated timber

Cross-laminated timber consists of layers of wedge-dovetailed softwood boards 17 or 27 mm thick glued to each other crosswise. The crossing effect makes the elements very stable in terms of form and they are thus suitable for wall construction. The covering layers can also be made of other timber products to improve the surface.

Depending on the covering layer and the number of layers, the panel thicknesses lie between 51 and 297 mm. The maximum width is 4.8 m. The units are made to a maximum length of 20 m to suit the particular purpose. Walls can therefore be built up to four storeys high using this material.

The values given in Table 11 apply to loads of up to 3 kN/m².

Tab. 11:
Standard values for dimensioning cross-laminated timber

Span	Unit height
3.8 m	115 mm
4.6 m	142 mm
6.4 m	189 mm

All the three solid ceiling systems described are also used as solid material for walls, following a similar principle known as <u>solid panel construction</u>. The number of new products on the market in addition the systems described is increasing constantly. The principle of wall construction is similar to that of modern log construction. The solid loadbearing wall is given heat insulation on the outside and a weatherproof shell.

ROOFS

Most masonry buildings include some timber construction. Pitched roofs usually have timber roof trusses, above all in detached house construction. But under close consideration, a timber roof truss on a masonry building represents a hybrid building method combining two different systems, lightweight construction and solid construction, a dry building method and masonry construction using wet mortar.

Pitched roofs

In timber structures, the pitched roof and the wall are constructed following a common principle.

Layers

The functions and sequence of the layers, starting from the outside, <u>weatherproof shell</u> with substructure, <u>sealing</u> or windproofing, <u>insulating</u>

> \\Hint:
> Here the reader is again referred to the volume *Basics Roof Construction* by Tanja Brotrück, in which the terms used below are explained.

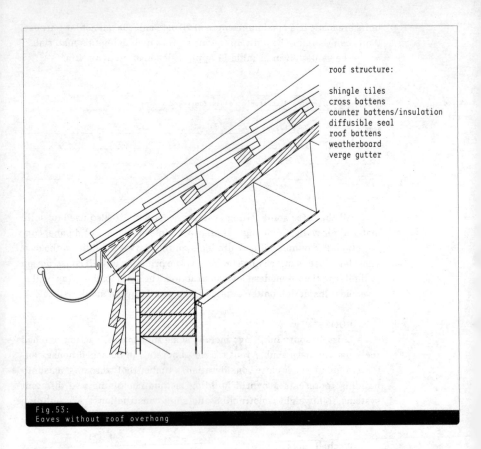

roof structure:

shingle tiles
cross battens
counter battens/insulation
diffusible seal
roof battens
weatherboard
verge gutter

Fig.53:
Eaves without roof overhang

layer, vapour barrier and internal cladding, are identical. But the aim is to join these layers at the point of transition from roof to wall to create a continuous envelope that fulfils all the scientific building functions without interruptions or weak points.

Anchorage

In addition, the roof loads from the roof structure's own weight plus snow loads in winter have to be transferred to the external wall. At the same time, the roof is anchored to the outside walls, to be able to absorb wind suction force, to which the roof is particularly susceptible. But as well as these numerous technical demands, design requirements must not be forgotten. The eaves and verge of a building make a particular impact on its architecture. A major issue is whether the roof should be constructed with or without an overhang.

Building science

The detailed section in Figure 53 shows a relatively simple solution to the transition from roof to wall, as all the layers can be merged

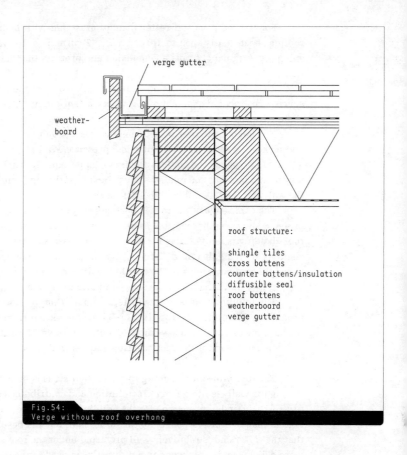

Fig.54:
Verge without roof overhang

without difficulty. The internal wallboard cladding on both sections is trimmed off at the edges at the transition point, but the mastic seal is scarcely perceptible in the room if the colour of the wall is matched to the joint.

Gluing the sheeting to the inside corner of the eaves and verge closes the vapour- and airtight envelope of the completed roof space.

The <u>insulation layers</u> between the loadbearing wall posts and the roof rafters are attached to each other directly. The header of the longitudinal wall and the gable wall form the upper conclusion.

No special purlin is required at the eaves as it is in masonry construction because the <u>rafters</u> can rest directly on the <u>timber frame wall</u>. Care should be taken at the verge to fill the gap between the gable wall and the peripheral rafter tightly with insulating material, to avoid creating a serious thermal bridge.

Outer skin For the roof, the full rafter insulation is clad with boarding and a roofing sheet that is open to diffusion ($S_d > 2$ m), and on the wall the stiffening plywood panel forms a conclusion and protects the thermal insulation.

The outer skin of roof and wall is ventilated from the rear. Each air space is entirely independent of the others, a ventilation system in its own right, with separate air in- and outlets.

Roof edge Not having the roof project emphasizes the volume of the building and makes it a more powerful physical presence. Wall and roof affect the overall impression the building makes almost equally. This impression may be enhanced by the related scale principle of the lap-joint cladding on the wall and the flat-tail shingle tiles on the roof.

The gable edge of the roof is particularly exposed to the wind, and is protected by a <u>weatherboard</u>. As screwing into the end-grain timber of the roof cladding is not permissible, the weatherboard is fastened to the roof cladding with a galvanized flat steel tie. Any precipitation appearing between the roof covering and the weatherboard is collected in a sheet metal <u>verge gutter</u> and directed to the eaves and into the roof guttering.

Roof projection The building makes a very different spatial impression with a roof projection. The projecting roof is more clearly detached from the body of the building and seems to have more of a life of its own. The different material language used for the roof covering and the wall cladding emphasizes the autonomy of both buildings

The roof projection running round all four sides is important in protecting the wall cladding from precipitation and contributes considerably to <u>structural timber protection</u> for the building. A corresponding disadvantage is that the rafters penetrate the outer skin. But the bevelling means that the rafter end is relatively well protected under the roof skin and the more delicate end-grain wood is not directly exposed to the weather.

> \\Hint:
> The gable wall, unlike all the other walls in the building, concludes not with a horizontal header, but with a header running diagonally at the angle of the roof pitch. The verge is cut orthogonally in the plan view to obviate distorted cross sections (see Fig. 54).

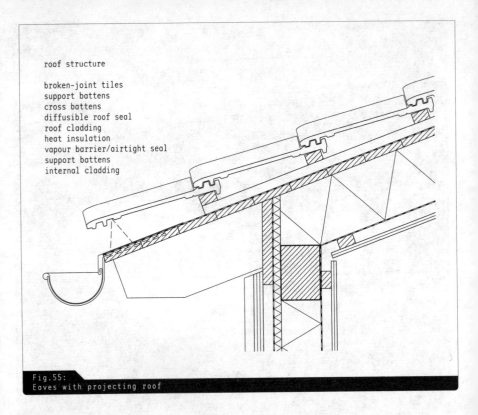

Fig.55:
Eaves with projecting roof

roof structure

broken-joint tiles
support battens
cross battens
diffusible roof seal
roof cladding
heat insulation
vapour barrier/airtight seal
support battens
internal cladding

Wall junction

In order to avoid laboriously matching the vertical board-and-batten cladding to the rafters, the cladding ends at their lower edge. A board is inserted between the rafters to cope with this; › Figs 55 and 56 it also fixes the upper edge of the cladding. <u>Ventilation</u> behind the external cladding finds an outlet via the air space between the inner board and the covering board.

Verge

At the verge, however, the wall cladding follows the edge of the roof, and extends to just under the roof cladding, where a gap of 2–3 cm is needed to ensure that air from the ventilation system can escape.

The roof projection on the gable side can be achieved only with protruding purlins. These support the outer rafter, called the <u>verge rafter</u>, which cannot be supported inside the building, but by the timber frame of the longitudinal wall. It runs out beyond the gable wall as an eaves purlin and thus continues to support the verge rafter. The framing timber should be dimensioned according to the width of the protrusion. It is better to replace the square cross section with an upright beam format.

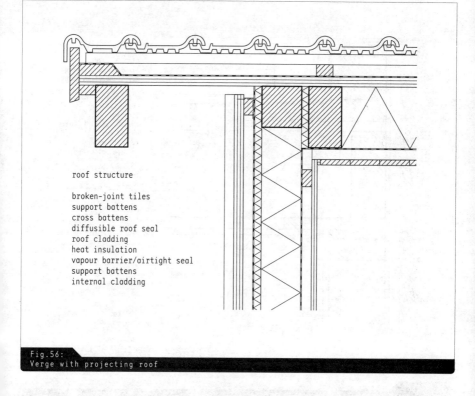

Fig.56:
Verge with projecting roof

roof structure

broken-joint tiles
support battens
cross battens
diffusible roof seal
roof cladding
heat insulation
vapour barrier/airtight seal
support battens
internal cladding

The verge detail shown in Figure 56 differs from flat shingle tiling as no <u>verge gutter</u> is needed. A specially shaped <u>verge tile</u> removes the water. It and the weatherboard conclude and protect the roof cladding. The weatherboard is screwed at the side to strips reinforcing the upper and lower sides of the edge of the cladding.

Flat roofs

Flat roofs are a key feature of modern architecture. They are generally used in the context of concrete structures. The flat roof is now also accepted as part of modern timber construction.

Parapet

In the detail shown in Figure 57 the external skin extends to the end of the roof as bevelled cladding, also sometimes called the parapet in the case of flat roofs. There is no need of a weatherboard to protect the roof skin, as used for the pitched roof gable, as this role is performed by the roof parapet, which protrudes at least 10 cm beyond the roof skin. This parapet and the open-top façade cladding are protected by sheet metal

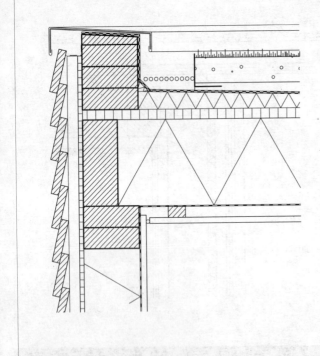

roof structure:

planted layer/
filter layer/drainage layer
root protection layer
roof seal
slanted insulation
vapour barrier
boarding
thermal insulation/beam
vapour barrier
support battens
internal cladding

Fig. 57:
Flat roof edge joisted ceiling – timber frame construction

cladding angled inwards. Air from the façade's ventilation must be able to escape at this point.

Seating

The roof ceiling is constructed and seated in the same way as the intermediate floors. › Fig. 49 The roof parapet, made up of two horizontal timber cross sections is reminiscent of the threshold running round the intermediate floors.

As in the external wall, the insulation is placed between the supporting joists of the roof. But as the rest of the structure is conceived as an <u>unventilated flat roof</u>, the vapour barrier on the inside of the roof insulation has a very particular part to play. An additional, inclined layer of insulation material is fixed to the roof cladding, to ensure a continuous slope on the firmly rooted roof seal down to the roof gullies. This is best achieved with piled granular material. This insulation later also bridges possible weak points in the roof insulation below.

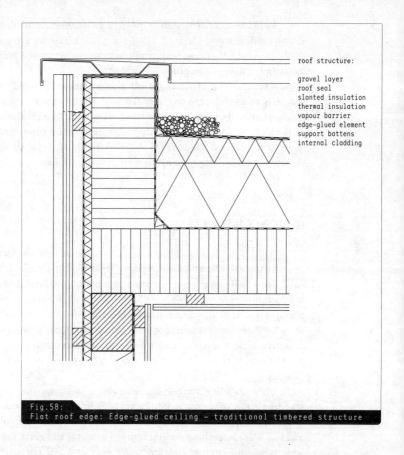

Fig.58:
Flat roof edge: Edge-glued ceiling – traditional timbered structure

roof structure:

gravel layer
roof seal
slanted insulation
thermal insulation
vapour barrier
edge-glued element
support battens
internal cladding

The low degree of storage afforded by a lightweight flat roof can be compensated for with a green roof. A strip of gravel at least 50 cm wide must separate the planting from the edge of the flat roof and the timber building components, for fire protection reasons.

The flat roof edges shown in Figures 57 and 58 each show only one of the many possible detailed solutions for the external design for the roof edge, as well as for the structure of the roof itself.

It should also be remembered that although an unventilated construction is a very common solution, flat roofs can also be constructed with ventilation, like pitched roofs.

A roof ceiling that uses edge-glued construction › chapter Ceilings is very close to a flat roof construction in solid reinforced concrete, in structure as well as in detail.

Solid wood construction helps to improve thermal insulation and the storage efficiency of the roof. The edge-glued elements are supported by the frame of the timbered wall, and the slope up to the roof parapet is also constructed using edge-glued elements.

A ventilated roof structure will be used above the solid wood ceiling, similarly to a reinforced concrete flat roof. Here, the vapour barrier is laid on the roof ceiling below the thermal insulation. The insulation layers of wall and roof are connected only indirectly via the solid timber components. The external thermal insulation layer in front of the loadbearing structure prevents thermal bridges from being formed.

IN CONCLUSION

At the end of our *Basics Timber Construction* volume it is fitting to emphasize timber's special qualities as a building material once again. In the Tectonics chapter of his book *Style*, Gottfried Semper called it the "primeval material of all stave constructions".

Timber construction means building elementally according to constructive logic and clear, easily understood laws. Differently from solid structures, the flow of forces in stave structures can be read and studied directly.

Understanding timber construction also opens the way to understanding many other construction systems used by architects. There is an immediate sense of parallels with steel construction, with joints of bars and surfaces. Something similar applies to metal and glass façades, which have taken over timber construction's post and rail approach. And even concrete, as a cast material, draws on the tectonic principles of timber construction for its loadbearing systems made up of columns, beams and ribs. The loadbearing effect of a reinforced concrete ceiling can be more readily understood if the reinforcing role of the loadbearing steel bars is translated mentally into timber beams.

Many architecture courses start with timber construction, for this reason. For example, it helps with understanding fundamental construction principles and is a multifaceted and unique field in which to work.

APPENDIX

STANDARDS

Timber construction generally

DIN EN 338	Timber structures – strength classes
DIN EN 384	Structural timber – determination of characteristic values of mechanical properties and density
DIN V ENV 1995-1-1	Eurocode 5: design of timber structures; Part 1-1: general rules and rules for building
AS 1684.1-3 1999	Residential timber - framed construction - Design criteria
AS 1720.1 1997	Timber structures – Design methods

Timber as a building material

DIN EN 338	Loadbearing construction timber – strength classes
DIN EN 384	Loadbearing construction timber – definition of characteristic strength, rigidity and bulk density values
DIN EN 1912	Loadbearing construction timber – strength classes – classification of visual sorting classes and timber types

Timber protection

DIN EN 335	Durability of wood and derived materials; definition of hazard classes of biological attack
DIN EN 350	Durability of wood and wood-based products
DIN EN 351	Durability of wood and wood-based products – preservative-treated solid wood

US Standards

Uniform building code, UBC

UBC V, Chapter 25 Wood

Handbook to the Uniform Building Code Part V Capter 25 Wood – An illustrative commentary

Wood – Frame House Construction, United States Department of Agriculture, Forest Service

Wood Handbook, United States Department of Agriculture, Forest Service

LITERATURE

American Institute of Timber Construction (AITC): *Timber Construction Manual*, John Wiley & Sons, 2004

Werner Blaser: *Holz-Haus. Maisons de bois. Wood Houses*, Wepf, Basel 1980

Francis D.K. Ching: *Building Construction illustrated*, 3rd edition, John Wiley & Sons, 2004

Andrea Deplazes (ed.): *Constructing Architecture*, Birkhäuser, Basel 2005

Keith F. Faherty, Thomas G. Williamson: *Wood Engineering and Construction Handbook*, McGraw-Hill Professional, 1998

Manfred Hegger, Volker Auch-Schwelk, Matthias Fuchs, Thorsten Rosenkranz: *Construction Materials Manual*, Birkhäuser, Basel 2006

Thomas Herzog, Michael Volz, Julius Natterer, Wolfgang Winter, Roland Schweizer: *Timber Construction Manual*, Birkhäuser, Basel 2003

Theodor Hugues, Ludwig Steiger, Johann Weber: *Timber Construction*, Birkhäuser, Basel 2004

Wolfgang Ruske: *Timber Construction for Trade, Industry, Administration*, Birkhäuser, Basel 2004

William P. Spence: *Residential Framing*, Sterling Publishing Co., New York 1993

Anton Steurer: *Developments in Timber Engineering*, Birkhäuser, Basel 2006

PICTURE CREDITS

Figures 11, 15, 18, 24	Theodor Hugues
Figure 14	Ludwig Steiger
Figures 23	Jörg Weber
Figure 36	Anja Riedl
Figure 39	Anja Riedl/Jörg Rehm
Figure 48	Ludwig Steiger/Jörg Weber
Figure page 12	Johann Weber
Figures pages 28, 67	Jörg Weber
Figure page 87	Architekturbüro Fischer + Steiger
All drawings	Florian Müller

导言

在1937年建筑师培训的一篇短文上,密斯·范·德·罗曾经说过:"任何地方的房屋或者高楼结构都没有古代木建筑表现得那么清晰,任何地方的建筑材料、建筑体系和形式都没有木结构表现得那么明朗!这是数代人对木结构的赞美。木建筑强烈地表现出木材的质感,散发出温和的气息,外观相当美丽!这听起来就像古老的赞歌。"这是一位20世纪最伟大的建筑师对木结构的魅力和其在建筑领域的地位强有力的赞美。

木材料种类繁多,可以采用的建筑结构体系也非常广泛。如果在学生的设计中采用这些材料,合理选择木结构构件和处理相应节点就需要丰富的知识。

不像学生所熟悉的整体式结构施工方式,木结构是由组装工人按照严格的安装程序,在定义好的结构格栅上进行组装的。根据设计要求,需要更多系统的方法以及大量细致的绘图工作。本书分三个阶段向学生介绍木结构。首先介绍木材和木材特性,然后介绍木结构建筑体系和它们的节点特征,最后介绍组装构件和组装方式。所有内容基于理解起来简单、处理起来容易的建筑思路,并适用于处理任一木结构的重要问题。理论上适合木结构的大规模承重体系,如桥梁或大厅结构没有着重描述,但是提供了广泛的阅读资料。

尽管木结构中存在一个特殊的难点,但是此难点也可以看作一个较好的机会。木结构施工技术不断发展变化,为了弥补现有的传统技术缺陷,已经逐步引进了大量的新材料和新技术。

本书的目的是说明木结构适用于广泛的领域,并对木结构作了简单的概述。虽然只包括传统知识的初步内容和经验结构,但是至少会对于新型建筑材料和新技术的发展产生有益的启示。

建筑材料

木材

全世界可以用作建材的树木有数百种。它们有不同的外形，性质也各异。大多数树木加工后用在家具制作上，只有数量较少的松类树木用于木结构建筑上。所以，对于木建筑的初学者不必是木材专家，重要的是真正理解它的内部结构，熟悉这种材料的基本物理特性。

木材的成长

在利用木材时，首要的是了解这块木料，或者这道梁、这块板是树木植物组织的哪一部分，以及该树木所处环境对它生长和质量的影响。世界上没有完全相同的两块木材，木材的特性首先取决于该树木的种类，其次是它来源于树木躯干何处位置。

树木的躯干是由纵向的管状细胞组成，在树木的成长过程中，这种细胞主要是负责营养的传输，细胞壁内填充着纤维素和木素（物质填充物）。细胞壁结构和细胞的框架决定了木材的强度。不同于素混凝土或砌块等建筑材料，木材是具有方向性的材料，与营养传递方向一致，从树干到树枝相对应。

细胞围绕着树干中心生长，树干中心为髓心，是树干的最初形成部分。树木通常以年轮的形式生长，温带区域树木每年的生长期从4月份到9月份，在此过程中生长成为树木的一个年轮。

在树木一个年轮中，春季的树木细胞腔较大，形成早期的软质木材；秋季树木的细胞壁比较厚实，形成晚期的坚硬木材。晚材成分的比例多少基本上决定了木材的强度。

> **提示：**
> 木结构建筑的承载力特性是指荷载作用在木材纤维的横截面方向上，横纹或者顺纹的方向决定承载力大小。因此，设计中必须包含关于安装方向的信息。在剖面图上，木材是横纹还是顺纹，必须表达清楚。

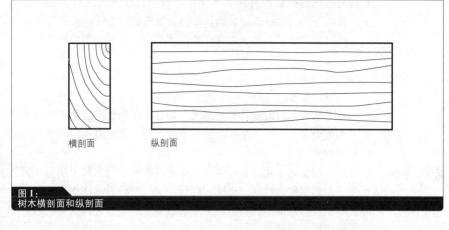

图 1:
树木横剖面和纵剖面

边材, 心材

通过观测树干的横截面, 就能够了解到树木的生长过程。根据树木此特点, 外侧<u>边材</u>与内部早期形成的<u>心材</u>有或多或少的不同。由于心材没有提供输送养料的作用, 因此相对于具有供输送养料功能的边材部分, 显得更为干燥。根据心材和边材的区别, 可以将木材分为:

— 心材树种;
— 较密实组织树种;
— 边材树种。

心材树种中, 树心颜色较暗, 边材颜色较亮, 区别明显, 具有独特耐气候性, 如橡树、落叶松、松树和胡桃树。<u>较密实组织</u>树种,

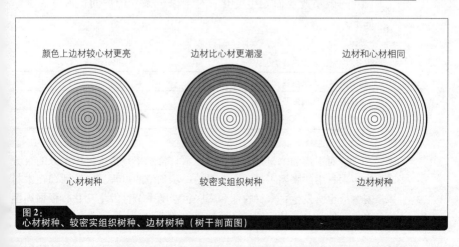

图 2:
心材树种、较密实组织树种、边材树种（树干剖面图）

心材与边材的颜色没有什么区别，都较亮，惟一的区别是在含水率上，树心干燥，边材潮湿，例如云杉、枞树、山毛榉树和枫树等。

<u>边材树种</u>中，心材和边材在颜色和水分含量方面都没有差别，主要包括桦树、桤木和白杨树。

P13

木材的潮湿

几乎所有木材的物理特性都会受到含水率的影响。木材含水率决定了木材的重量、耐火性和防虫、承载能力，甚至更重要的木材空间稳定性和坚固性。

收缩，膨胀

随着湿度的变化，木材会发生膨胀和收缩。当木材干燥时，其体积变小，通常称为收缩；相反的过程，潮湿时将引起体积增大，通常称为膨胀，因为细胞腔和细胞壁中含有水分。作为一种吸水材料，木材能够根据周围环境而释放或者吸收水分。这种功能也被认为是木材的胀缩。

建筑木材的水分含量必须详细说明，下表给出定义：

成长期的木材	水分含量高于30%
半干燥的木材	水分含量高于20%，低于30%
干燥的木材	水分含量低于20%

建筑木材应该在干燥的状态下进行安装，如果不能，安装场所的湿度最好能控制在一个相对固定的水平上。木材平衡含水率随着木材尺寸的微小改变而发生变化。对于以下条件的房间，木材含水率如下：

封闭房屋，有供热装置	9% ±3%
封闭房屋，无供热装置	12% ±3%
上侧覆盖，四周敞开的房屋	15% ±3%
敞露房屋	18% ±6%

绝对干燥的木材是不存在的，木材含水率是指木材中水分的百分率。木材中的胀缩不是一个简单的一步到位的过程。胀缩活动在木料安置后也会发生；木材收缩和膨胀也会随季节性发生变化，根据周围大气湿度，冬天的胀缩活动要比夏天的少些。

切割类型

由于树木边材和心材之间的水分含量不同，以及木材内部的年轮早材和晚材不同、收缩比率不同，因此，木材的切割方式也非常复杂，关键因素是所用木材处在树干的什么位置上。

木材可以<u>切向切割</u>，也可以<u>径向切割</u>，即在年轮中以合适角度切割，其角度大小决定了木材体积大小。根据木材类型，切向切割的木材比径向切割的木材的收缩度要高两倍多，纵向收缩度忽略不计。

体积变化也意味着从树干直角切割下来的厚板或者方木会发生的不同扭曲。由于年轮的缩小，横切的厚木板远离树心的一侧会发生弯曲。只有中心板材，髓心板材才能保持直线型，尽管它的边材较薄。图4表示出木材切割后的体积收缩（绿色）。

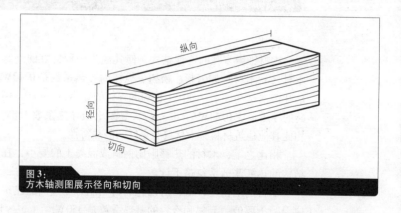

图3:
方木轴测图展示径向和切向

提示：
木结构施工一条重要的原则是木材的安装必须考虑由于木材收缩和膨胀引起的缩胀现象，即要求在相衔接的木料之间留有足够的空间。理想的一块木板只用一个螺丝固定中间部位或者一侧的边缘部位，确保木材能够顺纤维方向移动（见"外墙"章节）

提示：
远离树心切割下木材的一面指定为左侧面，含有树心的一面是右侧面。作为建筑使用的木材应当考虑预先的变形。

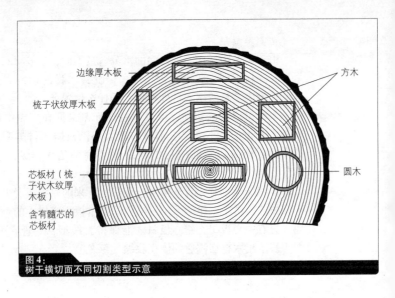

图4:
树干横切面不同切割类型示意

P16

特性

木材细微多孔的这种结构,使其成为一种良好的绝缘材料。松类木材(软木材)如云杉、松树、枞树的导热系数是 $0.13W/(m·K)$,落叶树木材(硬木)如山毛榉树和岑树的导热系数是 $0.23W/(m·K)$。砖的导热系数 $0.44W/(m·K)$,混凝土的导热系数 $1.8W/(m·K)$。相比其他建筑材料,木材具有良好的隔热特性。

相比之下,木材的热膨胀比钢材或混凝土的要小,在建筑中,木结构的热膨胀通常忽略不计。

毛密度

木材的毛密度较低,所以它的蓄热能力比坚固的建筑材料如砌块或混凝土要低。云杉和冷杉的蓄热系数是 $350Wh/(m^3·K)$,混凝土的标准蓄热系数是 $660Wh/(m^3·K)$。对于夏天的热防护,这个问题尤为显著。凉爽的夜晚和暖和的白天之间的热补偿,木材不如实体结构。较低的毛密度也意味着木材的隔声系数较低,但是木材细胞开孔,吸声较好。

蓄热能力和隔声效果只有采用重量较大的建筑材料才可以实现,即这些建筑材料需具有较大毛密度,如墙体采用石膏板或是纤维水泥,或者采用较重的楼面料。

防火

尽管木材是一种易燃材料(正常是易燃的),但是它在火中的燃烧性并不像起初看到的问题那么严重。具有较大横截面的木材,由于木材在燃烧时积累了大量的木炭层,因此从外到里燃烧均匀而且较慢。

> **提示：**
> 毛密度表明某种建筑材料的力度指标，它取决于材料的重量，以 kg/m³ 来表示。欧洲软木材毛密度是 450kg/m³ ~ 600kg/m³，硬木材是 700kg/m³，欧洲海外硬木材能够达到 1000kg/m³。相比之下，标准混凝土毛密度在 2000kg/m³ ~ 2800kg/m³ 范围。

> **提示：**
> 在德国，耐火性分为 F30B、F60B、F90B 等级，表明每种规格的木材在火灾中 30、60、90min 内保持其承载能力。

要完全失去承载能力需要持续一定的时间。这与钢铁结构不同，钢结构虽然不易燃，但是在高温条件下容易变形而失去它的承载能力。

湿度较高的木材燃烧率较低，横切木材的纹理，软木材的燃烧速度大约在 0.6mm/min ~ 0.8mm/min，橡树大约是 0.4mm/min。另外，木材燃烧性也取决于外部形状。同样容积的木材表面面积越大，耐火性就越低。对于实木，存在收缩裂纹燃烧率非常明显的状况。因此，没有裂缝的胶合板耐火时间更长，计算燃烧时间就比实木精确。

所以采用适当的尺寸，木材就能够满足防火的要求。

承载力

不像砌块是理想受压荷载材料，木材受压和受拉的承载力相等。但是上述的木材是管状细胞组成，荷载方向至关重要。木材能够承担的压力荷载，平行纹理方向（即沿着纵轴的顺纹抗拉）将近四倍于横纹方向抗压能力。横纹抗拉相比顺纹抗拉更为显著。根据德国标准，图 5 列出针叶木（S10）的允许强度（N/mm²）。

对于木结构的安装，应尽可能地有效利用木材的顺纹方向，来承担压力和拉力。

通常，承载能力取决于木材细胞壁厚的比例，即木材的密度。硬落叶木材如橡木特别适合用于承受压力荷载，如作为窗台或门槛；长纤维的松类木材，则更适合承受弯曲荷载。

树木自然成长外形不规则，作为建筑材料，首先它的预期承载能力就难以保证。因此，根据可以确定的特性诸如树枝的数量和大小、纤维的交错、裂缝、毛密度和弹性模量等参数，从视觉和机械上进行归类，评定销售等级。

在德国，建筑木材的承载力分为三类或三等，静力计算的木材强度分类见表 1。

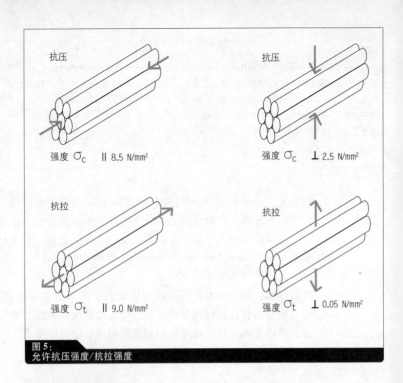

图5：
允许抗压强度/抗拉强度

表1：
德国的分类和等级

分类	等级	承载能力
S13	Ⅰ	高于平均水平
S10	Ⅱ	一般
S7	Ⅲ	较低

其他国家的木材分类标准可能更加复杂。在美国，所有的建筑木材都贴上标签，说明下列信息：等级，质量检查机构，锯木厂出厂编号，木材类型，水分含量，弹性模量 E，弯曲强度和用途。

木材运输到建筑工地更容易建筑监理管理。

木结构产品

木结构产品首先介绍实木；然后介绍木产品，其木材料的初始结构具有显著的变化；最后介绍建筑木板，该板是木材和其他材料如水泥和石膏粘结组成。

P20　　　　实木

实木包括剥去树皮的原木、切割的软木或硬木。在锯木厂，从树干上截取加工成特殊截面和长度要求的建筑木材。根据厚宽比，木材可以分为条板、木板、厚板和方木，见表2所示。

尺寸　　对于当前的条板、木板和厚板的厚度和宽度尺寸见表3，其他的尺寸体系区别只有微小的不同，相差仅几毫米而已。

建筑木材通常是用锯加工，没有刨平。木板和厚板两边为满足视觉要求刨光，每侧需要预留约2.5mm扣除，1.5~6m长度的木材需要预留25~30cm。

树干的方木包括方形和矩形横截面的木材，整个尺寸都可以利用。常用的方木截面见表4。

美国的树干方木是以英寸（1英寸约25.4mm）为单位的，最小宽度2英寸，较小的横截面主要是适合用于木框架结构的支撑位置。（见"木框架结构"和表5）

中欧和北欧最普通的木材类型包括云杉、冷杉、落叶松和花旗松。在美国包括花旗松、铅笔柏、卡罗琳娜松和北美脂松。

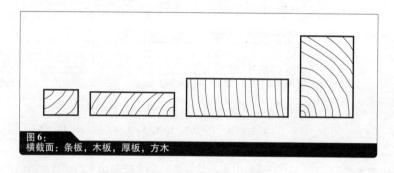

图6：
横截面：条板，木板，厚板，方木

表2：
条板、木板、厚板、方木的横截面

	厚度 t	宽度 w
	高度 h [mm]	[mm]
条板	$t \leqslant 40$	$w < 80$
木板	$t \leqslant 40$	$w \geqslant 80$
厚板	$t > 40$	$w > 3d$
方形板	$w \leqslant h \leqslant 3w$	$w > 40$

105

表3:
常用的木材横截面（单位：mm）

条板横截面	24/48, 30/50, 40/60
木板的厚度	16, 18, 22, 24, 28, 38
厚板的厚度	44, 48, 50, 63, 70, 75
木板/厚板的宽度	80, 100, 115, 120, 125, 140, 150, 160, 175

表4:
常用方木的尺寸（单位：cm）

6/6, 6/8, 6/12, 6/14, 6/16, 6/18

8/8, 8/10, 8/12, 8/16, 8/18

10/10, 10/12, 10/20, 10/22, 10/24

12/12, 12/14, 12/16, 12/20, 12/22

14/14, 14/16, 14/20

16/16, 16/18, 16/20

18/22, 18/24

20/20, 20/24, 20/26

表5:
美国木材的尺寸（单位：英寸）：

宽度	2, 2.5, 3, 3.5, 4, 4.5
高度	2, 3, 4, 5, 6, 8, 10, 12, 14, 16

实木产品

　　下面的实木产品是指对实木精细加工和油漆之后的产品。

　　建筑实木（SCT）通常以强度等级进行分类，也根据其外观形状进行特殊分类。木材应满足承载力、外形、尺寸和正稳定性、水分含量、最小裂缝宽度和表面质量等指标要求。楔端节点，是指在木料的端部通过榫木连接，这样可以制作任何长度的木料，常用在方木上。

　　用二合梁或三合梁同样也能提高实木质量，指两块或三块木板或方木沿顺纹方向粘结在一起。

　　十字横梁是由四分木顺纹面粘在一起的。这里每块圆木外侧转向内侧，在矩形截面中心处形成贯穿整个梁长的管孔。

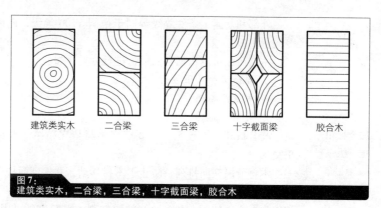

图7:
建筑类实木,二合梁,三合梁,十字截面梁,胶合木

胶合多层木材（胶合木）能够满足严格的承载力和稳定性要求。它是由多层软木板在外压作用下,沿顺纹方向压制而成,采用含有苯酚、间苯二酚、三聚氰胺或者聚氨酯的人造树脂胶粘剂来防水。粘结处外部颜色从深棕到浅棕各异。

木板在粘结和刨平前必须保持干燥,木料上的任何的瑕疵都需要机械去除。薄板粘结意味木材横切面上相邻处没有任何变形。胶合木常用于跨度较大的承重结构,因为它的横截面可达200cm,长度可达50m之长。

木产品

木产品是一类特殊的经济型木材,它们由木材加工中的废弃品如刨花、纤维等和木材组件如木板、板块、单板和单板条等再利用而成。

通过人造树脂胶粘剂或者矿业胶粘剂压制的工业化生产过程,原产品性能将得到大幅度提高,木材不一的质量也会变得统一。相比实木,木制产品的静力特性和抗拉力更加精确,收缩膨胀特性也比实木要低。

木材产品通常以标准尺寸的面板形式提供,例如125cm宽的面板。

而全世界的木制产品根据粘结方法来分类,因为可以提供该产品在潮湿条件下的使用性能。德国的分类如表6所示。

美国木产品有四个粘结等级,见表7所示。

外部和暴露1条件等级等同于德国V100等级，而内部条件等级等同于V20。

木材产品可以根据自然成份进行分类，如：
— 胶合板和层压板；
— 木屑产品；
— 纤维产品。

胶合板和层压板

胶合板和层压板至少需要三层木材依次粘结，木材的纹理依次正交。

木材分层交向排列，防止了木材移动，并保证面板各个方向具有必要的强度和稳定性。因此，胶合板和层压板特别适用于加固木结构和作为承重墙使用。利用适当的胶粘剂，它们也可以用于外部工程，其边缘部位容易受到潮湿的影响，如果采用覆层或封闭就可以作为外墙板使用。

表6：
德国的木产品等级划分：

V20	不适用于潮湿条件下
V100	适用于短时间内暴露在潮湿条件下
V100G	适用于长期暴露于潮湿中，防止真菌

表7：
美国木产品的等级

外部的	持续暴露于潮湿中
暴露1	雨期中具有高抵抗性，不适用于长期暴露
暴露2	正常地暴露于潮湿中
内部的	保护内部，不应暴露于潮湿中

注释：

为了确保面板开发利用的经济性，建筑格栅必须以面板尺寸确定，尽可能减少浪费。对于宽度为125cm的面板，轴线距离可以为面板的1/2长度即62.5cm，或者1/3长度即41.6cm，能够达到经济用木的目的。

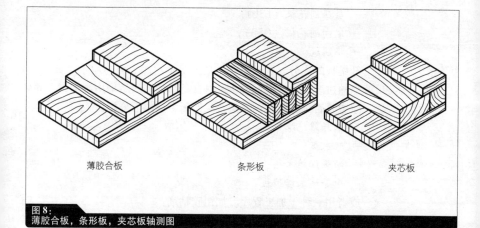

| 薄胶合板 | 条形板 | 夹芯板 |

图8：
薄胶合板，条形板，夹芯板轴测图

薄胶合板根据木板的厚度而定（8～33mm），由三层、五层、七层或九层彼此粘结制成。不少于五层，并且厚度大于12mm的胶合板，通常也称做复合板。

条形板和夹芯板，如众所周知的芯板材，至少用三层粘结而成，中间层的条形板材交叉于所覆盖的胶合板，其板材具有良好的承载力特性。

最常用的胶合板和层压板包括：
— 薄胶合板；
— 条形板和夹芯板；
— 三合板和五合板；
— 层压板；
— 复合板。

木屑产品　　木屑板是利用木工业的废弃产品制成。木屑板由锯屑和刨花压缩加工并胶粘而成。不像胶合板是连续的板层，木屑板内部结构方向不定。可被用作加固（见"加固"章节），如用于加固墙壁、地板、顶棚和屋顶，或者当作楼板的干性抹平层。

定向刨花板　　建筑业常用的一种芯木板叫定向刨花板，之所以这样命名，是由于它是相对较长（大约是35mm×75mm）的矩形木屑或刨花组合成的定向结构。这种方向逐层交替变化，具有像胶合板一样的力学特性。这意味着它们有更高的强度，能够达到普通木屑板的2～3倍。

常用的木屑产品如下：
— 压制平板；

— 层压刨花板（LSL）；
— 定向刨花板（OSB）；
— 挤压板。

纤维板

纤维板的成分（颗粒）比木屑产品的更小。因为原木（针叶木）尺寸缩减过多，已经很难辨认木材自身结构。产品采用湿加工，并依据工序中是否施加胶粘剂、是否施加压力作如下分类：

— 木纤维绝缘板；
— 软木板；
— 沥青纤维木板；
— 中等硬木纤维板。

或者用一种<u>干加工</u>程序，加用胶粘剂：

— 中等密度纤维板；
— 高密度纤维板；
— 硬木纤维板。

通过湿加工而成的木板是<u>软木板</u>，因此经常用于内部结构、隔声、隔热以及屋顶定模。<u>中密度纤维板</u>由于它的同质结构，是流行的家具制造和内部装修材料。<u>高密度</u>和<u>硬纤维木板</u>主要用于立面覆层。

结构木板

不像前面介绍的组织性粘结木产品，而是非有机结合式的木板产品叫做结构木板。初始材料只包含一些木材的成分，或者根本就没有。它们一般被分为<u>水泥结合板</u>：

— 水泥木屑板；
— 水泥纤维板。

和<u>石膏结合板</u>：

— 石膏木屑板；
— 石膏板；
— 纤维板。

<u>水泥结合板</u>具有较高的防水、防寒特征，能够抗菌和防虫。因此，它们通常大量用于面材和与地表接触的基础部分。在木结构建筑中，它们是一种能力非常好的板面材料，也适合用于加固。

另一方面，<u>石膏结合板</u>只适用于内部结构，石膏板主要用于墙壁和顶棚的面层，而且几层纤维木板通常用作楼板的抹灰层。如果石膏纤维板采取了对抗天气影响的永久保护措施，可用于外部墙板。

P27 害虫

木材防护

不像矿石材料如砌块和混凝土,木材是一种有机材料,容易受到植物类的侵害(真菌类)和动物的侵害(昆虫类)。一旦受到侵害,木材的外部结构将会受到影响,进而削弱其承载能力,甚至破坏整个建筑结构。因此在木结构中,木材的防护就成为至关重要的一点。

<u>真菌</u>的生长需要纤维素。它们在潮湿、温暖和不通风的环境下生长得更为繁茂。木材的含水量一旦达到20%,就可以导致真菌腐烂。

<u>昆虫</u>,这里主要指甲虫,经常用松类树木的边材部分寄养幼虫。白蚁也是对木材危害性最大的昆虫之一。它们主要生长在热带和亚热带地区,美国和南欧地中海国家也能发现它们的踪迹。白蚁对木材的破坏,从木材外部很难发现,因为它们在木材内部挖掘出一个通道系统,以防止水分的流失。大批孳生白蚁的建筑物或者家具,当其承受不了外部载荷时就会突然倒塌。

众所周知,木材防护就是要在木材已经受到害虫侵袭的情况下加以<u>控制</u>,或者是进行<u>预防性保护</u>,以确保其不受害虫侵袭。

设计木建筑时,木材防护是最重要的。有三种必要的措施可资利用:

— 精选木材;

— 结构性防护;

— 化学性防护。

P27

精选木材

建筑木材要求:干燥良好、合理储藏。木材的含水率不超过20%。

许多国家的分类也详细列出了哪些类型的木材具有自然抵御昆虫侵害的能力。这些包括心材如柚木、樟木、红铁木、橡树和洋槐等树木(见"木材的成长"章节)。在美国,此类型树木包括刺槐、黑胡桃树和红杉树。这些木材可以不需要化学防护措施,直接用于那些木结构暴露于潮湿环境的地方。但是,这些木材如果与土壤接触,就必须使用化学性防护。

P28

结构性防护

在结构性防护中,设计者起着关键作用。设计,特别是细部设计,应当考虑避免木材和木结构构件永久性的潮湿渗透影响(见"构件"一章对于基础、窗户和屋顶的处理措施)。

化学性防护

只有在其他防护方法失效的情况,才采用化学性保护。

化学木材防腐剂包括水溶性防腐剂、油溶性防腐剂和油类防腐剂。为了避免环境污染,木材防腐剂应该在封闭的设施中使用,例如通过蒸汽压力处理或者通过木槽注入等方式进行。只有切割表面和钻孔才在建筑施工现场进行。

化学性木材防腐剂要根据建筑构件的功能来施用,共分三类:

— 承重和加固部位;
— 非承重部位,并且尺寸变化;
— 门窗部位,尺寸稳定的非承重部位。

对于承重部位,必须采取预防性保护措施。当地规程决定是否采用化学性防护,即根据风险程度和相应风险等级而决定。

当下列条件下和采用具有自然抵御能力的木材类型时,可以不采用化学保护措施。

门和窗户

窗户和外门属于非承重体系,但尺寸要求严格,为使其功能完好,所容许的误差非常小。需要特别的防护方法避免其潮湿。如采用保持长久强度的心材,也可以不用化学制剂处理防护。

非承重构件且尺寸要求不太严格的,误差容许范围可适当放宽,其中包括相互搭接的覆层,栅栏和棚架。在没有防腐处理和油漆的情况下,依然可以进行建造,只要业主能够接受日久构件将越变越灰的状况。化学处理后的木材在任何情况下都不可以大面积地应用在室内。

注释:

不仅在木结构建筑方面,在其他特殊部位,也应注意以下方面:
1. 远离潮湿(屋顶凸起,凹陷水滴板);
2. 确保水顺利导出(倾斜水平面);
3. 保证潮湿部位空气流通(通风空间)。

提示:

接触土壤的木材部分腐蚀性更大。任何结构都应该避免木材接触土壤和永久性暴露于潮湿中。木结构典型的做法:使木材远离地面大约30cm的距离。

木结构

瑞士建筑师保罗·阿塔利亚曾经写道："木结构建筑必定要建造，而石结构将会没落"。这种有点偏激的说法并不是有意否定石结构建筑的合理性，而是要表明木结构建筑的特殊合理性，即基于杆系结构和木结构结点体系。

结构稳定性

对于如何满足结构稳定性的基本要求，本书提供了一些方法或是建筑体系，即通过承重体系和相邻构件的节点连接。

在介绍完木结构的重要结构后，再接着介绍木结构静力承载要求。

承重结构

建筑物的结构稳定性取决于多种因素。首先，所使用的建筑材料必须有足够的承载力要求，合适的尺寸能够承担竖向荷载，如墙体、屋顶和顶棚。地基也必须能够承受这样的荷载。

所有的建筑物都会受到水平作用力，主要是来自风荷载和冲击荷载，会影响结构的水平稳定。

加固

抵抗水平作用力的一种方法是加约束或刚性连接，如图10所示。木支撑固定在基础上，具有抗弯刚度，阻止它们向侧面移动或者发生变形。最简单、最基本的约束方式是将支撑木材的削尖一端撞击到地基里。钢筋混凝土结构中，约束混凝土柱的筒式基础已经被普遍接受。但这种连接方式在木材的防护方面成了问题。

> 提示：
> 这部分内容可以详细参考本套丛书中的《承重结构》，中国建筑工业出版社2010年2月出版。征订号：18857。

斜撑

对墙体和顶棚适当加固，能够使木结构成为一个三维刚性结构。设想一个纸板盒，它的周边墙壁相当容易被推成菱形，除非加上盖子。即在第三个方向加上水平盖子，才能使其成为一个稳定的结构，这就是加固体系，如图9所示。

加固的基本单元是固定三角形。在木结构中，这种三角形的形式部件固定在一起，相对容易实现。三角形支撑结构就是常见的一种形式，能够使得矩形的墙体框架成为一个牢固的板面。

图10表示了几种加固平面结构的方法：a) 用两个受压构件，根据外部两个方向的作用力，交替承担荷载；b) 用一个拉/压构件，根据外荷载方向变化，承担拉力和压力；c) 钢索只能承载张力。根据外荷载方向变化，拉索交替承担拉力。

三角形受力这种效果，在木结构中能够通过平面单元来实现，加斜撑板或铺设厚板来达到加固。

这些方法在木结构设计中应用得非常普遍。在下面所讨论的木结构建筑中，关于墙体如何加固的方法有一定的区别。古建筑中木框架结构大多采用斜撑，十字形的钢索形式是现代木结构的一个特征。

木结构体系

建筑方法根据不同的外部条件而定：气候、文化特征、可用材料、工具以及技术水平等。在木材贫乏的南欧地区，石建筑比较繁荣；然而木材丰富的北部地区，木建筑比较发达。但是，木建筑也有地区之别。在阿尔卑斯和德国中部的多山地区，原木建筑采用实木做墙体，同样，在北欧用大量笔直的针叶松。相比之下，在中欧和东欧地区，落叶木生长繁荣，传统的木结构建筑比较流行。

> 提示：
> 一个平面结构如墙体或者顶棚，如果能承担纵向轴力并且没有变形，就认定为稳定的平面内结构。如果能够承担较小横向作用力，产生横向变形，就认定为平面外结构。根据所受外力的方向决定平面结构是平面内结构还是平面外结构（见图9）。

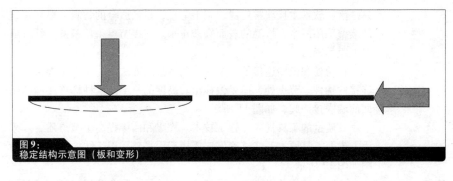

图 9：
稳定结构示意图（板和变形）

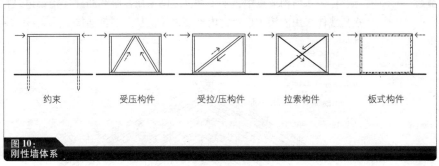

约束　　　受压构件　　　受拉/压构件　　　拉索构件　　　板式构件

图 10：
刚性墙体系

19 世纪，特别是在 20 世纪，新兴技术和新型材料在很大程度上改变了欧洲的木建筑。建筑工程发展了节点连接技术，高质量钢结构节点能够更好地处理木结构横截面问题。这些技术在<u>骨架建筑</u>中大量采用。在北美，木框架结构采用简单的铆钉节点建立了<u>肋拱式建筑</u>。

木材行业的大市场上不断涌现出新型天然和合成材料，新的运输方法和隔热需求的提高推动了木结构建筑的不断发展。

直到今天，木结构建筑史就是从原木到木结构、从木框架到骨架结构的杆系单元建筑原理的故事。因此，从事木结构领域的任何人都非常有必要熟悉和掌握以下所述的系统，其中主要包含板材的新系统，以此扩展木建筑的未来。

P34

原木结构

术语"编织建筑"（strickbau）也经常见诸专业文献，因为梁末端交叉连接，所以被描述成编织在一起。

原木结构的一个特征是需要大量的木材，水平搁置的构件十分容易跌落。软木生长非常规律，躯干笔直，是最适合的材料。最初建筑的墙壁采用原木，接触的表面轻微正平。连接节点采用苔藓、纤维或者毛织品密封。

拐角处和横墙连接处，木材采用嵌接的方法，通常采用榫来紧密连接两构件。节点处，横梁偏移出其高度的一半。这种榫连接能够使两面墙嵌接一起，如图11所示。

高质量的工具提高了施工技术。榫眼和凸榫提高了节点质量。采用方木而不是原木，能够确保墙壁横截面平整。现代结构中梁的剖面图如图12所示。

节点

传统木结构节点中，水平杆件之间采用榫眼和凸榫连接；建筑物拐角墙体采用嵌接头连接；内外墙采用抗拉燕尾榫连接。（参见152页的图45）

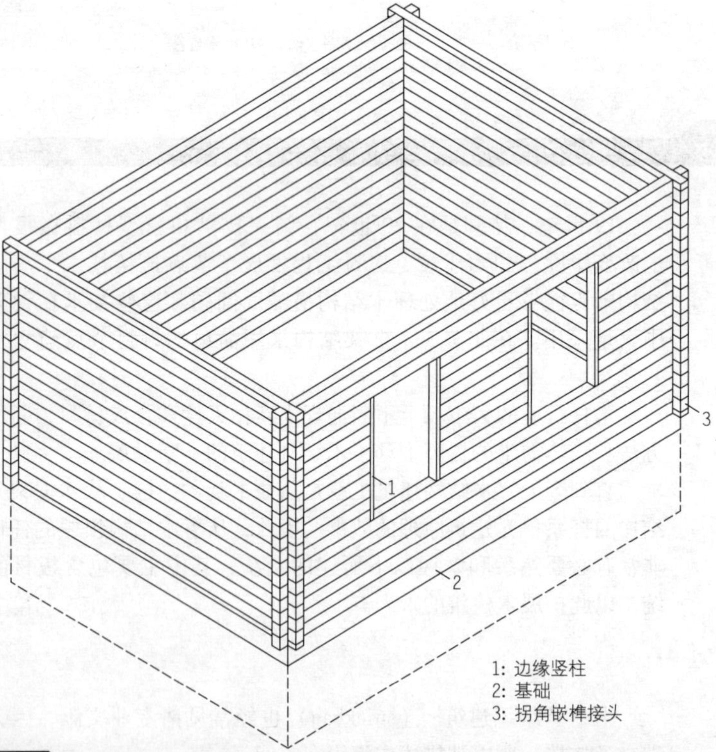

1：边缘竖柱
2：基础
3：拐角嵌榫接头

图11：
原木结构轴测图

116

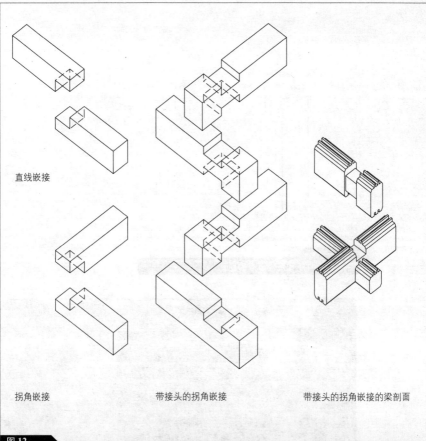

直线嵌接

拐角嵌接　　　　　带接头的拐角嵌接　　　　带接头的拐角嵌接的梁剖面

图 12：
嵌接图，带有嵌齿的拐角嵌接轴测图

墙体横截面由多根横梁组成，已经不能满足现代保温的需要。因此，现代的原木结构需要另外设置隔热部分。理想的隔热装置在墙体

> **提示：**
> 　　一个嵌接就是一个节点，这个节点木材凹槽厚度为木材的一半，紧紧地组合在一起。能够承担起作为外侧齿榫产生凹凸榫接的拉力。同样，这也可以应用于圆锥形的楔形榫头上。

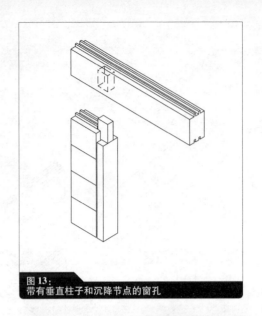

图13：
带有垂直柱子和沉降节点的窗孔

图14：
原木建筑——外部拐角，内墙连接，窗户缝隙

外侧，避免冷缩。传统原木结构的水平外侧铺装木板，这能保护隔热装置不受天气影响。木行业现在提供的层压原木结构墙，这种类似三明治的墙体由厚板和绝缘部件组成。

沉降　　　　原木结构中，梁横纹方向承受挤压荷载，产生较大的沉降，每层楼可以达到2～4cm。因此，制作门窗的时候必须充分考虑这个因素。框架的垂直部分或者柱子顶端应切槽充分，能够承受墙体的沉降，不产生二次弯曲，见图13所示。窗框或门楣也通过隐蔽结合点来处理沉降。同样的原因，任何垂直贯穿建筑物的烟囱或者设备不应该固结于建筑物上，而应保证它们可以活动的方式固定。所以虽然许多原木建筑物看上去很简单，但是实际上需要大量的技术和经验才能够完成。

原木结构施工需要有严格的矩形地面设计安排。

正面设计应当确保任何孔隙尽可能地小或少，这样墙体结构就不会存在较大的损害。原木结构本质上是实体结构体系，对于所熟悉的砌体结构开孔技术，原木结构也可以适当采用，如图14所示。

P38

传统木结构

传统的木结构清晰地展现了结构上的荷载传递。因此，德国专业出版物上有时称它是"Stil der Konstruktion"（建筑风格）。对于结构木材和墙体填充部分之间的承重和非承重结构看起来差异非常明显，所以此类建筑具有特别强大的吸引力。

填充

在各个承重柱子之间的空间叫做栅格或是间隔。历史上，木结构

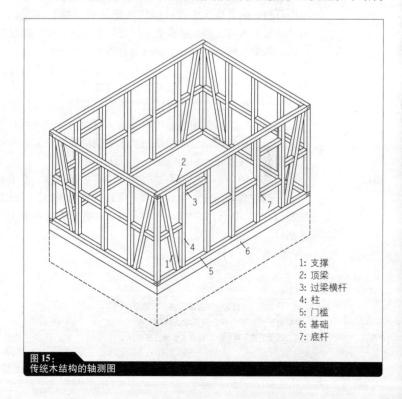

1：支撑
2：顶梁
3：过梁横杆
4：柱
5：门槛
6：基础
7：底杆

图15：
传统木结构的轴测图

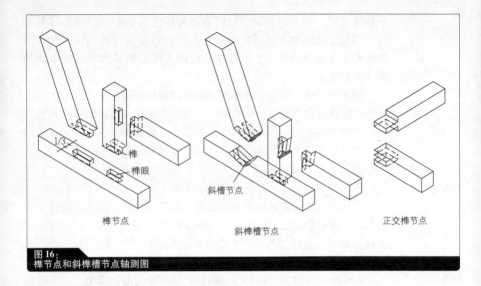

图 16：
榫节点和斜榫槽节点轴测图

采用石料，或者是黏土和枝编物（抹灰篱笆墙）填充。现代，出于内部保暖需求的提高，基本上采用绝热填充物，外部的覆层也有保护作用，隔绝气候影响。有时，内部也需要一定的覆盖物。

　　填充物是不能承重的，它很有可能是用于填补木墙壁面板上的孔隙。窗户不能随意安置，但是如果符合建筑柱网要求，也可以大量使用。因此，传统木结构要比原木结构更容易为房间提供充足的光线。

节点

　　典型的木结构常用榫和榫眼连接，直线向连接木材。斜榫槽节点可更好地传递载荷，也经常使用。水平的门槛和顶梁也采用榫节点或者采用拐角嵌接（如图 16 所示和"原木结构"章节）。

横截面

　　木结构的一个典型特点是竖柱、横杆以及斜撑，在下侧被门槛固定，上侧被顶梁固定。木材所承受荷载主要为压力，标准尺寸的方木横截面：10/10cm，12/12cm，14/14cm。矩形横截面常用于门槛和顶梁部分。

> 提示：
> 榫节点的木材是被分割成三块的凹陷部分，这样有利于榫眼和凸榫合理连接。榫眼不超过 4cm 深，这样主要承重杆件横截面不会被过多削弱。

120

木结构分一层或多层。水平木材仅用于门槛和顶梁，沉降度可以大大降低；这是原木结构比不了的。

顶棚　　顶棚横梁搁置在顶梁上，对于没有覆层的木结构，可以看到整个梁架的布置方式，上一层的木结构在顶棚横梁上建造，以门槛部分开始。两层或更多层木结构的承重结构从尾顶贯穿到底部，顶棚必须在两个墙壁之间悬挂起来。（如157页的图50所示）

历史上，木结构多采用硬木材，如橡木。各个地区的建筑方法也不同。例如在德国，弗朗科尼亚、阿莱曼尼克和萨克森的木结构就有明显的差异。木结构构件也因地区不同而各异，也用不同的名称。

栅格　　柱子之间的最小距离通常是100~120cm。然而，历史上的木结构建筑也有更小和更大的间距。尽管受到各种结构的限制，但是木结构建筑在施工和设计方面仍然开发了大量的创造可能性。纪念碑保护权威机构试图保护那些形式多样的历史性木结构建筑以及维持它们在城镇风景中的地位。因此，当利用现存的树干材料建筑时，建筑师应熟悉传统木建筑的原理。在现代的木框架结构中，大量的柱子、支撑、横梁等引出节点处绝热材料的填充问题，造成了大量的工作量。现在，高水平的手工艺处理木材节点工序已经大量由电脑轧制所取代。

P40　　木框架结构

现代的木框架结构起源于北美。沿着新铁路线安置下来的村镇需要简单、经济的建筑，以便短时间内建成。木材正好能满足这种需求，适用大陆的不同气候条件。

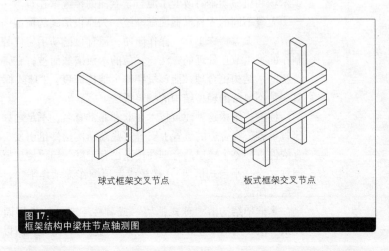

球式框架交叉节点　　　板式框架交叉节点

图17：
框架结构中梁柱节点轴测图

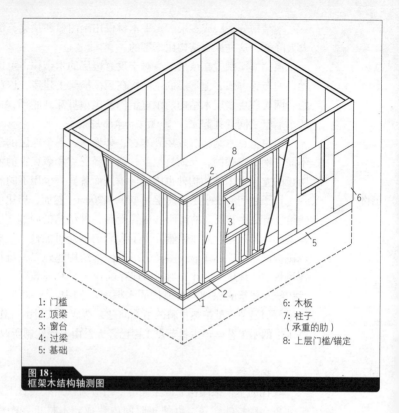

1: 门槛
2: 顶梁
3: 窗台
4: 过梁
5: 基础
6: 木板
7: 柱子
（承重的肋）
8: 上层门槛/锚定

图 18：
框架木结构轴测图

在 19 世纪的前 50 年，工业技术开始影响到木结构的建造。蒸汽锯木场和机械切割钉改变了源自于欧洲的传统木建筑。

大量不同的木材横截面渐渐统一为厚板形式的横截面。简单的钉合节点，无需特殊工具，操作简便，适宜自己动手，代替了需要精心制作的手工节点处理的方法。轻薄的木板横截面被钉在一起，构架比传统的木结构更合理，能够贯穿整个结构高度。"**球式框架**"起初是对一种不寻常的轻型结构的嘲笑说法。

<u>柱梁式结构</u>这种建筑技术在欧洲非常有名，不足之处是获取木材比较艰难，在合适的位置上安置这些高结构构件也引发一些难度，垂直构件贯穿整个楼层的合理性传递也有难度，导致楼房层状叠加，美国称之为"**平台式框架**"。楼板或者顶棚作为工作平台，在框架结构中组装而成。

<u>木框架结构</u>由"球式框架"或者柱梁式结构发展而来，是一种建筑方法，当框架还在地上的时候用墙壁进行组装。所以框架通常只

有一层高，尽管也能见到一些两层木框架结构的实例。

肋骨结构　　所有这些建筑技术有时被统称为肋骨结构，特别是在德国（Rippenbau），这些垂直的部件连接紧密，厚板的横截面非常窄。

横截面　　20世纪早期的欧洲建筑主要集中于将混凝土作为建筑材料。直到20世纪80年代，木构架建筑才被广泛地接受，此时，美国木结构被用来寻求一种更合理更有价值的建筑研究方法。美国标准"二乘四"英寸在欧洲变成6cm×12cm，而且因此在某种程度上生产出了比美国尺寸更坚硬的木材横截面，美国的标准大约是5cm×10cm。

节点　　对于木框架结构，典型木材节点采用一种钉对接方式，对角钉能够在横向接缝上产生最大强度的横纹向的联接。只有在嵌板加衬制作时，才采用刚性节点连接，防止钉被拔出。(见图19)

栅格　　木框架结构带有小栅格密肋间距，通常匹配面板的宽度以便于嵌板。通用单位宽度为62.5cm（见"木产品"章节）。确定建筑格网后，绝缘材料的标准宽度也就成了决定性的因素。

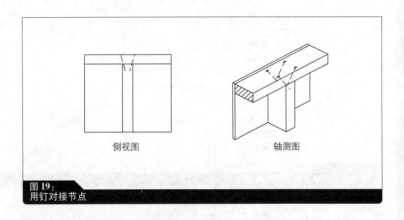

侧视图　　　　　　　　　　轴测图

图19:
用钉对接节点

提示：
　　对于木建筑"二乘四"（英寸）截面是一个经过试验测试的横截面，而且能够用于多种方式组合。有时候"二乘四"在木框架结构中是一种通称。今天，欧洲中部气候条件下，更好的绝缘标准通常意味着更厚的横截面，比率达到"二乘六"。

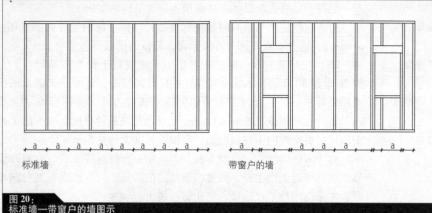

| 标准墙 | 带窗户的墙 |

图20：
标准墙—带窗户的墙图示

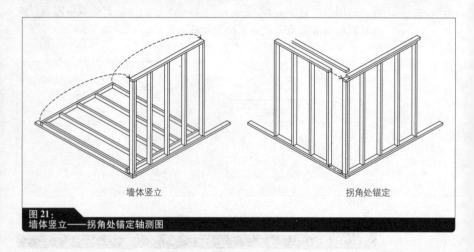

| 墙体竖立 | 拐角处锚定 |

图21：
墙体竖立——拐角处锚定轴测图

  木框架结构的一个典型特征，是建筑的总长度并不一定与单位间距的整数倍相一致。结构栅格的重复模式（见图20）在墙端没有严格要求，而是以一个特殊的单位尺寸终止。窗户的安排和内墙尾端也可以自由处理，它们的位置由设计决定，而非结构栅格，这是传统木建筑的处理办法。框架式木结构的结构栅格的基本目的就是经济合理地利用材料，而不是从建筑和审美的角度来考虑。因此，在与其他木建筑体系相比之下，对于建筑物的平面图或者剖面图，设计几乎没有任何限制和约束。

装配 装配阶段，这些结构木料不像传统木结构那样竖立起来再进行组合，而是在地面上钉成框架，然后在门槛上竖立起来，门槛是用钉子牢牢地锚定在地板上的。这种由门槛和木框架组成的双重门槛是框架式木结构的一种典型特征——利用侧边而不是复杂的木节点连接，换句话说，在必要的位置甚至可以用双倍或三倍的截面来加厚。（见图21）

那些用这种方法建立起来的墙体是基于同样的原理锚固的。墙体上部分的框架和上层木框架组装起来：上层的门槛通过外围圈梁，从一片墙横跨到另一片墙，外围圈梁拉紧结构。

顶棚　　下一步放置在木框架墙上的横梁也是由符合顶棚跨度的薄截面木材构成的。顶棚四周环绕同样高度的横梁，通过钉子固定，可以预防倾斜的发生。顶棚铺装完木板后，梁的接合点才完全稳定。这样，顶棚横梁形成了一个刚性平面，就能用作下一楼层的平台，二层的墙壁用相同的建筑程序建造。（见图21）在美国，通过平台式框架，可以

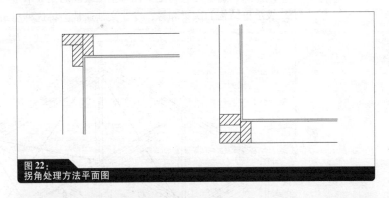

图22：
拐角处理方法平面图

图23：
木框架结构墙体——内部和外部

125

建造出 6 层高的楼房。

> **提示：**
> 固定墙壁时必须布置角柱，以便在两个方向安装半个横截面宽度如 3cm 的内部覆层时有可靠的支撑，解决方法见图 22。

P46

木骨架结构

木骨架结构是从木结构发展而来的，因为现代需要更自由的空间分割，需要大面积的玻璃安装。一般而言，木结构中偶尔会用到骨架式结构。

大体上，如果建筑物是由柱和主梁组成的一级承重结构来支撑一个二级承重结构的次梁和椽，专业文献通常会将这种建筑方法定义为

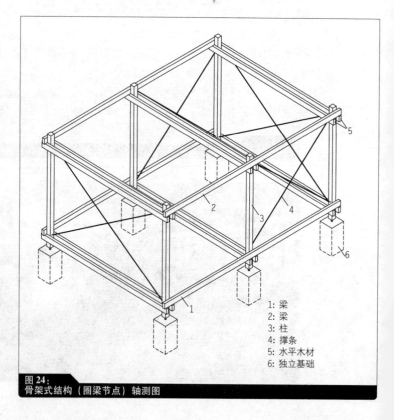

1：梁
2：梁
3：柱
4：撑条
5：水平木材
6：独立基础

图 24：
骨架式结构（圈梁节点）轴测图

骨架式建造。墙体组成后形成房间，墙体承担各自骨架的荷载。这就赋予了大面积安装玻璃的可能性，以及平面图设计更大的灵活性。木骨架结构实现了 20 世纪现代主义者"表皮和骨架"的原理。

栅格　　承重柱之间的距离通常比木框架结构明显要大，也比传统木结构中的柱距大。承重骨架通常内外均可以看到。胶合层压板在木骨架结构中发挥着重要的作用——胶合层压板形成的梁，能够满足较大的支撑空间（见"木结构产品"章节）。

　　风荷载通常由交叉的钢绳索或者圆钢柱承担，它们仅仅承受张力（见"加固"章节）。

　　过宽的柱距意味着应采用独立基础。柱子通常独立于墙体，支撑的基础不带覆层，所以带有柱脚的镀锌钢材连接点在木骨架结构中是一个明显的建筑细节。

　　多层建筑物，并不是一层一层建造，而是采用连续的柱子。附在两根柱子上的水平梁成为圈梁（见图 24 和图 25），或者横梁对接（见图 26 和图 27）。

节点　　在木骨架结构中，柱子和梁利用金属装置连接，没有显著削弱木材横截面。不像原木和传统木结构中的手工艺连接节点的尺寸由结构

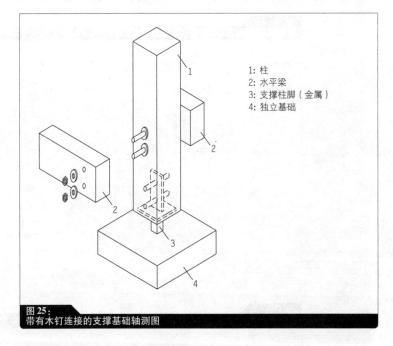

1：柱
2：水平梁
3：支撑柱脚（金属）
4：独立基础

图 25：
带有木钉连接的支撑基础轴测图

127

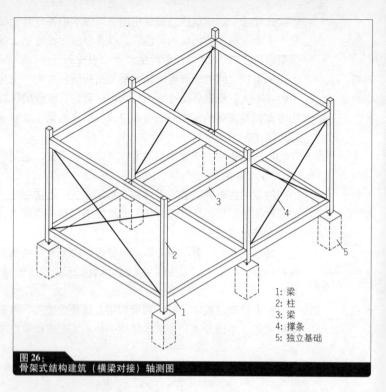

1: 梁
2: 柱
3: 梁
4: 撑条
5: 独立基础

图 26：
骨架式结构建筑（横梁对接）轴测图

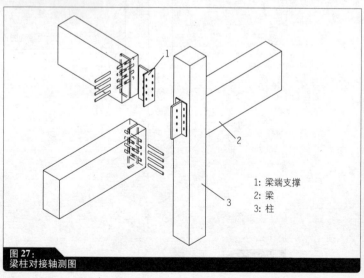

1: 梁端支撑
2: 梁
3: 柱

图 27：
梁柱对接轴测图

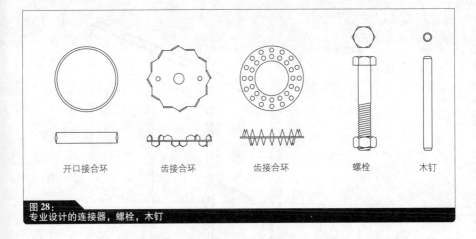

图 28：
专业设计的连接器，螺栓，木钉

（开口接合环　齿接合环　齿接合环　螺栓　木钉）

工程师根据施工技术表制定。因此，它们是以<u>工程连接</u>而出名。

连接材料　　木结构组件横向连接起来后，为了更好传递荷载（见图 25）、<u>专业设计的连接器</u>（见图 28），通常是被嵌入或敲入到一个环形物，承受着节点处的荷载，并且将其传递到最大数量的木材纤维上。这些节点用<u>螺栓</u>紧紧地锚固在一起。

另一种连接器是一种圆形的钢销子，钉入预先钻好的孔洞，传递榫入木材中连接钢板的荷载。对接节点，如梁柱之间的连接方法，从外形上基本看不出来。（见图 27，图 50 以及"顶棚"章节）。

P49　　　　　木面板结构

木结构的特殊优势包括相对较短的组装周期和干性的施工体系，这就意味着完整的结构可以马上被利用。预先在工作间制作的单独部件、墙体单元甚至整个房间单元更缩短了建筑周期。

预制件　　木框架结构中的框架式墙非常适合在木匠工作间预制。尽量将木结构从施工现场转移到木匠工作间预制，能够使建筑过程避开天气环境的影响。

木面板结构可以使这种方式获得最大化。预制面板单元通常有一个完整的楼层高度，经过绝缘处理，并安装了与建筑部件相关的所有层以及内外侧的覆层，这样在工地只需要竖立起来并固定在一起就可以。一个关键的细节就是节点或单元缝。楼板和顶棚也是预制的，需要固定在墙体单元上或悬挂在两片墙之间。

作为板式建筑的木结构仍需要相对较大比例的手工操作。但是，

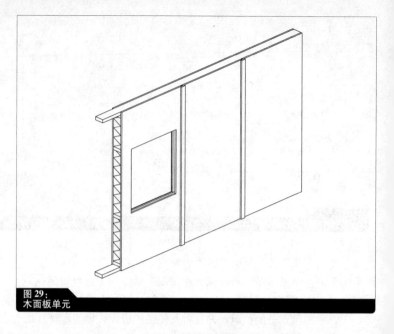

图 29：
木面板单元

最近几年有一种向工业化发展的趋势。随着为满足板式结构需求的木工业发展，承重墙体采用坚固的层积胶合板或者边缘胶合单元加工而成，差不多可被称为厚板结构。

这就意味着木结构逐渐从栅条向实体结构转变。因为数十年来，木结构所采用的原料尽可能地减少。现在看来，这种趋势似乎发生了转变，由于工业建筑方法和对原材料更加连续的利用所导致（见"木产品"）。

厚板结构

使用混凝土面板的厚木板结构在东欧国家和斯堪的纳维亚半岛国家十分常见，因为这种混凝土板非常沉重，单元尺寸受限于一层高度。相比之下，木材较轻，为避免单元高度可达 4 层之高。另外，按序建造就可以避免不适当的空间性危险，而这个通常是预制和统一立面单元的一个常见问题。这样的产品可以通过 CAD 和电脑控制的机器很容易地生产和制造。

这种厚板结构方法是否比栅条结构更为公众所接受，取决于各种各样的经济因素。对于栅条结构的以下劣势——精细和昂贵的升降设备、更大的输送难度和需要大型生产车间，需要向工人支付更高的工资。

构件

本书的第三部分着重介绍了关于基础、墙壁、顶棚和屋顶的详细要点,以及它们整体组装木结构建筑的方式。特别需要注意的是那些相对复杂的层状结构和相邻建筑构件的连接与节点。

像木框架结构和传统的木结构一样,建筑的每一构件都是以1:10的比例尺寸来举例详细描述的。不需要具体举证来源或试图以一概全,图例表现了相关文脉和产生的问题。

基础

由于木结构的防护要求,任何木结构的基础都应该高出地面30cm。木结构或者立足于地下室的顶棚上,或者没有设置地下室的情况下直接建于混凝土或者砌体基础之上。假如地下土层的天然地基不能够满足设计要求,那么木建筑可以采用以下三种类型的基础(见图30)。

— 板式基础　　＝　平板
— 条形基础　　＝　线性
— 独立基础　　＝　点性

板式基础

任何木结构体系都可以采用板式基础。板式基础特别适合用于木框架结构,因为它需要一个平台来组装施工。混凝土地板下要么铺设一层压实的抗冻砂砾(粗砂砾)层,要么在冻结线下建造抗冻墙。

> 提示:
> 这本书将只在必需的外部墙壁节点处,介绍屋顶建筑和在屋顶上的结构的处理,同本套基础教材中《屋顶结构》中提到的一样,中国建筑工业出版社2010年出版。

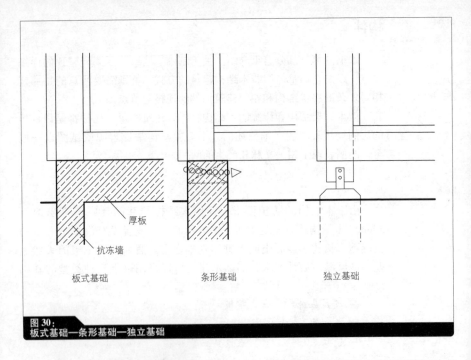

板式基础　　　　　条形基础　　　　　独立基础

图30：
板式基础—条形基础—独立基础

假如建筑物下面设有地下室，那么通常情况下地下室的顶棚采用混凝土，而不是厚木板。

P53　　　　　条形基础

如果建筑物采用条形基础，那么此建筑地板必须由木板制作。这里的关键是木结构的防护，地板结构下应保证通风。地面和木托梁之间应闭合，以防止小动物的进入，但应保证适当的横向通风来确保托梁周围的空气流动。

此类基础另一个要点是隔热处理，木材铺设在条形基础后，隔热层能够嵌固在上侧托梁之间。因此，首先木板必须固定在事先安装在托梁两侧的板条上（见图33）。另一种做法是安置预制隔热顶棚单元。

这两种情况下，根据条形基础之间的跨度大小，托梁的尺寸都和地板单元大小相近（见"顶棚"章节）。

P54　　　　　独立基础

独立基础比较合适木骨架结构。木骨架结构中，建筑物的荷载主要由很少的几根柱子承担并传递到地基上，这使得地基开挖土方量缩

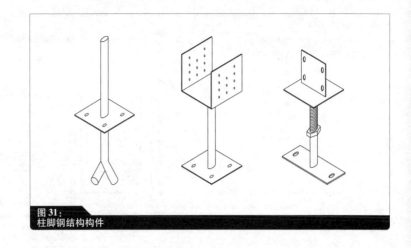

图31：
柱脚钢结构构件

减到最小。

对于木骨架结构，地板类似中间层楼板，铺设在主梁和次梁上。

为了满足木结构的防护要求，木料加工业生产出各种形状的钢结构连接构件，能够通过木柱与混凝土基础连接的节点将柱子荷载传递到基础上（见图31）。尽管已经这样处理，仍然需要特别注意一个关键问题——保持基础尽可能通风并保证意外发生的流水顺畅地流出（见"木材防护"章节）。

P54　　墙脚

不管何种建筑方法，对于建筑物墙脚区域都需要特殊的处理。墙脚容易受到地面潮湿、喷溅水流或者冬天雨雪的影响。出于对木结构的防护（见"木材防护"章节），外墙应该高出地面30cm，通过防潮材料与基础相连接。

在裸露的混凝土上建造基础是一个常见的解决办法。混凝土能够保护木结构避免潮湿的影响，并在建筑外观上给出了一个可见的结束。因此，混凝土板式基础一直延伸到外墙处。立面面板后采用浇筑的混凝土，可以建造出合理的通风道。为避免在基础表面做一些不必要的标记，木料覆层不用磨光，在滴水板边像有轻微的缩进。木料覆层后的通风区域外侧应设防虫网（见图32）。

木结构保护　　因为门槛外露并且接近潮湿地面，应采用抗潮湿的硬木来做。规划部门应该在本国木结构保护标准的允许下，研究不采用化学防护的可能性。木材和潮湿的混凝土之间必须存在防潮介质。板式基础上的

133

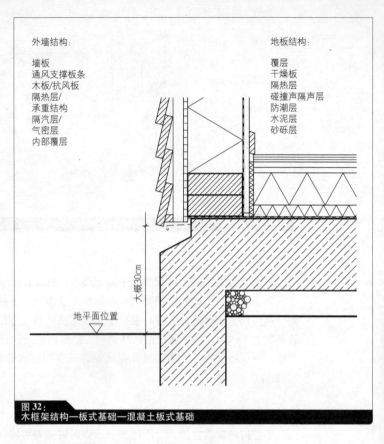

图32：
木框架结构—板式基础—混凝土板式基础

地板密封层和外墙上的隔汽膜起到了这样的作用。

图33展示了条形基础和外侧承重墙之间的处理方法。条形基础宽度能够保证承重墙体的门槛和地板的托梁。不需要设置防潮层，但是木地板和地面之间的空隙必须保证能永久通风。通风栅栏或者砖块应该和基础连成整体，确保建筑物四向通风。

> 提示：
> 木结构建筑锚固在基础上。门槛和建筑物柱子应该和板式基础或条形基础连成整体，通过规则的连接铆钉或者承重型木钉间断连接。

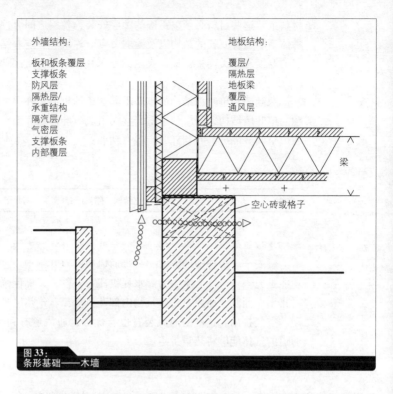

**图33：
条形基础——木墙**

建筑物的可见基础是决定木结构外形的一个关键因素。如果像上述的例子那样，降低建筑物的入口高度给人感觉建筑像没有基础一样，可以降低与建筑物紧紧相邻的地面高度。建筑物周边沟渠略低处设置基础，并采用木结构防护措施。考虑到安全问题，最好还是用一个铁格栅将沟渠覆盖。

P57

外墙

作为建筑物的保护层，外墙必须承受一定的强度。外墙经常遭受风雨的侵袭，也受到温度波动、噪声和放射性物质的影响。从内到外，外墙必须满足温度变化、空气对流、声音传播和蒸汽的扩散这些要求。

P57

层式结构

砌体结构以及传统的原木结构中，同一种材料承担着承重、隔热、密封和保护作用。轻型的木骨架结构将这些功能分配在不同层的

135

材料上。这些材料必须以正确的顺序安装，各层之间合理搭接，但这种体系不适合冗长的结构以及薄弱节点。建筑师决定着结构形式、各层的厚度和分类连接。最初决定的是采用单叶墙还是双叶墙结构（见图34）。

双叶墙和单叶墙结构的本质不同在于承重墙和外层之间的通风问题。单叶墙结构，外层和承重墙共同组成一叶；双叶墙结构，墙体被通风空间分隔成内叶和外叶，各有功能：

外叶	防风雨，通风
内叶	防风，隔热/承重结构，隔汽层/气密层，密封层，内侧覆层

建造科学

通风

最安全、最普通的方式，是带通风层的双叶结构。这个通风层的功能是充当一个膨胀室，为渗水提供压力补偿，以确保临时流水的及时排出。同时，通风道将建筑物内部蒸汽排出，绝热层外的湿气由流动的空气排出。这样的通风装置也有优势，通风道有利于夏季隔热，通道的气体能够将热量带走。

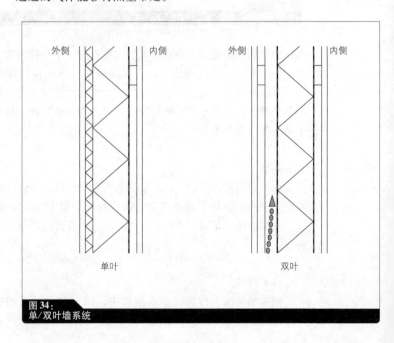

图34：
单/双叶墙系统

通风道必须至少埋深20cm，不能因为诸如窗户和基础等其他因素而减少。空气由基础处进入，从屋顶边缘和顶墙处排出。进口和出口处必须采用防虫网，防止昆虫进入。

防风　　防风材料用在隔热层外侧防止隔热层过快变冷；在隔热层和木结构的接缝外侧防止空气渗入。

在外露节点的外侧覆层（见"外部覆层"章节），防风材料也应具有保护隔热层免受湿气渗透的功能。为了满足内部蒸汽向外渗透的要求，防风层必须尽可能地允许蒸汽的扩散。木产品硬化板具有这样的功能，如果对接点损伤，这种材料外侧承重单元可替代硬化板的能力。否则就要采用薄膜或者护墙板。

隔热层/承重结构　　木结构建筑受到建筑师和广大客户的青睐，主要一个原因就是与砌体结构相比，具有更好的隔热性能，能够更好地储存热能。12～16cm厚的填充隔热材料木板，虽然内侧或外侧安装了隔热层，但是木板本身也能够提供很好的隔热性能。

市场上的大多数材料都可以用作隔热材料。但是，挤压延伸成型的聚苯乙烯（EPS和XPS）和所有的<u>刚性泡沫板</u>材料有些问题，因为它们和木材的伸缩率不匹配。在泡沫板中添加<u>隔热纤维材料</u>能够解决这个问题，因为纤维材料压缩性良好。

另外一个解决方法是使用再生纸制作的<u>松散纤维素隔热</u>。这种材料只适用于紧密结构的空间，通过吹制而成。这就使得它非常适用于由柱子和板块建造的闭合房间的框架式结构隔热。

纤维素隔热材料必须防止受潮，否则墙体就会体量增大并引起结构的破坏，难以修补。采用硼化合物可以防腐和防燃。

隔汽气密层　　"木材防护"一章详细地介绍了木材必须防潮。不仅应防止建筑

> **注释：**
> 承重结构和隔热层之间的空隙，可以在结构层外侧或者隔热层内侧填充附加的隔热材料。低密度的木纤维面板（见"木产品"一章）经常被采用，因为这种材料具有良好的稳定性，可用于防风；经过沥青处理过的木纤维面板，能够很好地防止外侧潮湿。

物外部潮湿渗透，还应该防止内部的蒸汽，因为它能够引起结构的

收缩。

湿气通过扩散（蒸汽）或者对流（内部空气）渗透到建筑物的每个部分，因此，木结构中隔汽层非常关键，它能够防止由于湿气渗透而降低隔汽层表现的影响。

在双叶和单叶结构中都可以采用它。尽可能采用连接整齐、渗透少的密封层。单一层就能具有防汽、气密的功能，设置在隔热层内侧。

抗热扩散

理论上讲，应建造外墙，目的就是要阻止蒸汽扩散到构件内部，而且那些已经渗透到内部的蒸汽可以被排出到外部。因此，需要注意的是，当安装墙体材料时，抗热扩散能力（S_d）从内部到外部呈现逐渐减小的趋势。

定义区分（S_d）值如下：

可扩散	（S_d）＜2m
蒸汽缓滞扩散	（S_d）2～1500m
抗蒸汽扩散	（S_d）≥1500m

蒸汽缓滞扩散适合用于双叶结构建筑。通常它由一种特殊的纸或者薄膜制作而成，建筑物内部的蒸汽能够渗透散发到外部，由隔热材料控制多少，并由通风层逸出。

单叶不通风结构需要内层隔汽层，用以阻挡蒸汽从内到外的扩散。这种隔汽层一般都是由隔汽的条形塑料或者是金属薄膜构成的。

外部覆层

原木料或者木产品外层在木结构建筑中能够提供防风雨的作用。在美国等国家，覆层通常也用其他材料如金属或者是石膏制作而成。仅仅出于结构方面的原因，混合材料都应具有相同特性，尤其木材是易膨胀或收缩的材料。作为一种建筑材料，设计应表现出木材的质量、表现性、表面和纹理等各个方面。

外侧覆层提供了大量的设计可行性，包括覆层类型的选择、板宽、纹理方向、木材类型、表面处理和光泽度等。

子结构

承重板条的子结构是不可见的，但它仍然是外部覆层的基本部分。子结构类型由覆层是否可以通风换气和板条布置方向是竖向还是横向来决定，连附在承重结构上。支撑板条根据外部覆层木板厚度确定间距（见表8）。

交叉布置原则要求垂直覆层的承重板条水平布置。这里，也需要垂直的板条或者通风板条，以保证从底部到顶部的空气连续流通，因为水平板条将会起到一定的阻挡。对于水平布置覆层的建筑，应采用垂直的承重板条。

安装

螺钉、钉子和支架用于安装木板。钉子可能会对覆层的表面和内部结构有一定损害。螺钉安装相对来说就比较安全，而且更容易控制。

并不是每种情况都需要防腐蚀材料，但是木结构表面应采用不锈钢钉或不锈钢螺钉和镀锌钉来防锈。木结构的安装应考虑木材的收缩或膨胀率。如果覆层板交替搭接，如板和条板之间或覆层节点处，钉子或螺钉必须只穿透一层板面。覆层条板应该被固定在单层铺板上或者节点处。螺钉或钉子，不应穿透次要组件的木料端部纹理。

表8：
板条间距

木板厚度 [mm]	板条间距 [mm]
18	400 ~ 550
22	550 ~ 800
24	600 ~ 900
28	800 ~ 1050

提示：
这个组件层的 S_d 值表示的是该层的抗扩散能力，相当于厚度为多少的空气层所具有同样的抵抗能力。它以米为测量单位，是材料厚度（S）与其抗扩散能力 μ 的乘积。$S_d = \mu \times S$ (m)。S_d 值越大的组件层，汽密性能就越好。

提示：
正面覆层木材的特点和作用在本套基础教材的《建筑材料》一书中有更详细的介绍，中国建筑工业出版社2010年出版（征订号：18810）。

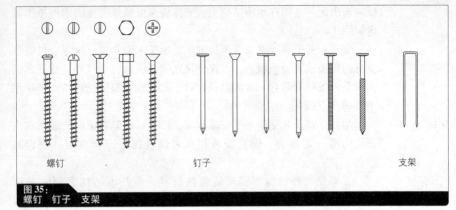

图35：
螺钉 钉子 支架

竖向覆层构件类型：(见图37)
— 板和板条覆层；
— 搭接条形板覆层；
— 盖板式覆层；
— 企口板覆层。

板和板条覆层　　板和板条覆层的底板和盖板之间搭接大约20mm。因此，如果用相同宽度的木板，就会产生盖板宽大和底板窄小的节奏感。这种类型的特点同搭接条形板覆层，侧表面相对坚固。

搭接条形板覆层　　搭接条形板覆层，相邻垂直覆层之间大约10mm的空隙由一块条形板覆盖，以防意外的水分渗入（见图36和图37）。

盖板式覆层　　盖板式覆层采用了磨光的条形板，像企口式覆层一样有相对光滑的外观。

注释：
　　板和板条覆层中，水平的板条足已成为支撑结构。覆层和内板之间不需要通风板条，其间的空气层能够垂直流通，并为外层表皮提供了适当的通风（见图36）。

注释：
　　板和板条覆层中，木板心材一侧向外（见"木材的潮湿"一章），当木板因为干燥而发生变形（弯曲）时，底板和覆板之间的节点依然保持闭合。

图 36：
垂直覆层：盖板式，搭接条形板，板和板条和企口式覆层

企口式覆层　　企口式覆层中，木板通过<u>回扣</u>或者舌榫连接在一起。这意味着可以将钉子藏在舌榫中或用金属钳夹来隐藏木板。连接木板之间留有适当的活动空间。对于企口板覆层，额外的木板或者条形板用来覆盖开口的角落。其他垂直覆层形式的角落通过双叶木板或者条形板封闭。（见图37）

水平覆层包括：（见图38）

— 搭接式覆层；

— 槽口式覆层；

和以下特定环境：

— 木瓦式覆层，

— 条形覆层，

— 面板覆层。

> 提示：
> 覆层单元之间的缝隙被覆盖叫做<u>封口式覆层</u>，比较而言，<u>开口式覆层</u>具有裸露的节点。对于开口覆层，防风层必须有防潮功能，防止隔热层的渗水。

141

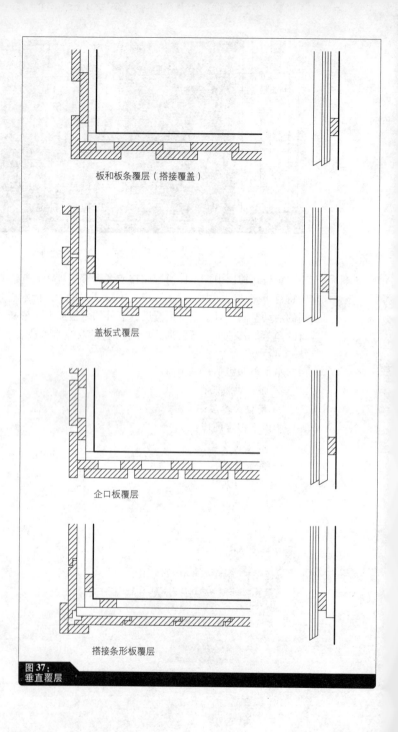

板和板条覆层（搭接覆盖）

盖板式覆层

企口板覆层

搭接条形板覆层

图37：
垂直覆层

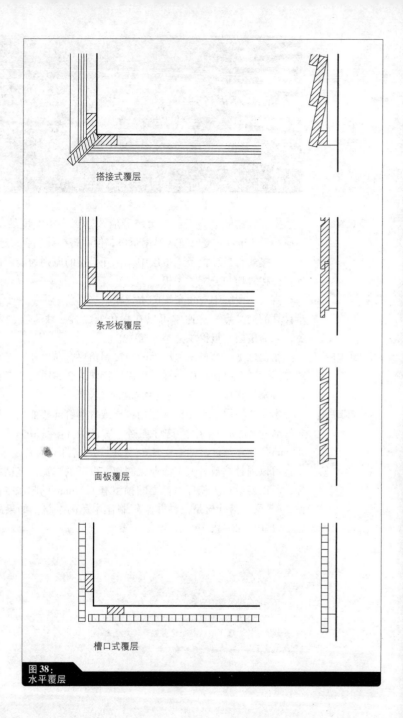

图38：
水平覆层

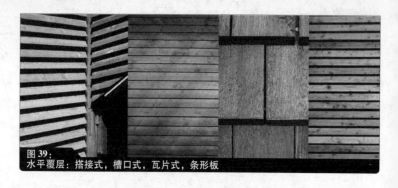

图39：
水平覆层：搭接式，槽口式，瓦片式，条形板

搭接式覆层

搭接式覆层中，相互搭接部分应该达到木板宽度的12%，并且不应小于10mm。覆板通常不用企口，但在搭接处用企口连接。

搭接式与板和板条式覆层中，相互搭接的原则为墙面以及与门窗各点交接时均匀分配了木板。

搭接式覆层中对角布置的木板在斜接拐角产生了一个几何学上很难计算的交叉点，这通常可以在拐角处放置一块垂直的木板加以避免，来为覆层木板提供完整性（见图38）。

槽口式覆层

相比之下，槽口式覆层上的斜接拐角更容易操作，隐蔽的固定方式亦然。同企口板覆层一样，可以用特殊的金属钳夹或者在舌榫上用钉子或螺钉连接来实现（见图38和图39）。

木瓦式覆层

木瓦式覆层由小规格木板用钉子或者螺钉连接而成，如同支撑板条的鳞片（见图39中右）。这种规格（宽度为50~350mm、长度120~800mm）的复杂性意味着它比较容易设计成曲形或者是柔和地变化。木瓦可以通过锯或者劈来商业化制作。劈开的木瓦细胞结构没有被破坏，所以寿命长。相邻木瓦之间预留有1~5mm的膨胀空间，通常铺设2~3层。落叶松是一种非常好制作木瓦的木材。如果屋面坡度比较陡（30°~40°），木瓦也能用于屋面覆层。

> 注释：
> 有关木结构保护的内容上（见"木材防护"章节），不像垂直覆层，水平覆层有一个优势，即基础部分损坏的木板可以被轻松地替换。这对于基础高度过低或者裸露基础的区域尤其重要。

条形板覆层　　条形板覆层是开口覆层形式的一种，因为缝隙没有被覆盖（见图38和39所示）通风空间对于渗透水的排泄非常重要。这种情况下，保护隔热层的防风层也必须起到防潮作用。对于木结构防护，用倾斜的条形板就非常有意义。条形覆层也可以垂直设置；斜角的条形板没必要采用。

面板覆层　　当采用面板覆层时，必须注意的是选择正确的材料，根据天气条件选择正确的胶粘等级（见"木产品"章节）。以下木产品适合应用于外部覆层：

— 胶合板；
— 松类木材的三合板；
— 水泥木屑板。

对于面板覆层，木产品的边缘是一个特殊问题。一个可靠的解决方法是利用覆盖条形板遮盖连接处。这可以保护易损的面板边缘。有些建筑师通过在面板上使用不同颜色的条形板来强调立面覆盖条形板的结构。

如果面板连接处是通过假缝结合，那么面板覆层平坦、光滑的效果就更加明显。在这种情况下，面板之间至少需要留有10mm的距离。面板边缘必须涂刷防水漆来防潮。

对于水平节点，面板的下部应该从下切割15°，这样可以确保任何水流都能够滴落下来。同时，必须确保任何水分都不可滞留在面板的上部边缘。另外，这些边缘必须被覆盖，最好是用金属构件，向外倾斜15°。（见图40）

面板通常用不锈钢螺钉固定。为隐藏安装痕迹，需使用吊挂面板

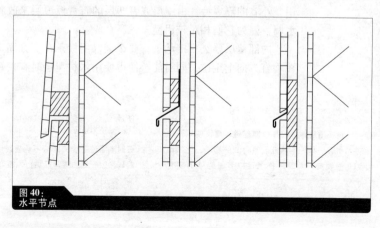

图40：
水平节点

的特殊子结构。这种体系非常有商业价值。

如果拐角如图 38 所示有对接节点，那么就需要注意开敞边缘不能面向季候风方向。

表面处理

外部覆层既不能承重，而且尺寸也不固定，没有必要采用化学性木材防腐处理。(见"木材防护"章节)

但是<u>物理性木材防护</u>是一个不错的选择。有目的地防止木材吸收过多渗透水，以及使表面免受紫外线的辐射。彩色的透明覆层或者不透明的油漆之类的物理性防护，能有效地防止木材随时间而碳化。

碳化

木材碳化只是颜色上的变化而已，木材本身没有被破坏。一些建筑师甚至故意将木材碳化效果作为一种设计效果。例如，MLTW（Moore, Lyndon, Turnbull & Whitaker）建筑师事务所在美国西海岸 (见图页 147) 设计的 Sea Ranch，木材表面经历了紫外线几十年照射和海洋气候的影响后产生了银灰的光泽，效果十分自然。设计细节时需要注意的是，在立面的凸出和凹进部分受天气影响的后果一样，因此可能会出现不那么想要的颜色。

色彩覆层

这里还有一种处理方法是喷刷色彩覆层。斯堪的纳维亚木建筑通常带有鲜艳的颜色，展示了这样处理的效果。

<u>透明色喷刷</u>或者<u>不透明色喷刷</u>都适用于木材上色。通常，油漆喷刷包括内层漆、中层漆和外层漆。关键因素是所用的产品彼此协调一致。

内部覆层和设备安装

木板覆层可作为内部覆层的一个选择，木产品如合板也可以采用。外墙的隔声特性由一层或者两层的石膏板覆层来改善，也叫做<u>清水墙</u>，因为它们相对较粗糙。

内部覆层要么直接承重，要么最好用板条在下方固定，更利于它在垂直方向上定位，同时能够使得使用木材覆层时有较好的通风。

提示：

面板覆层具有较高的敏感性，建议使用预防性木材防护措施，降低天气对其的影响，比如说建筑的朝向，大型悬挑屋顶或者是连续的阳台面板。

注释：

覆层木板在组装前应该上漆，这样当木材收缩时未经过处理的区域不会显现出来。两侧都应起码涂上一层油漆，以防止木板弯曲 (见"切割类型"章节)。

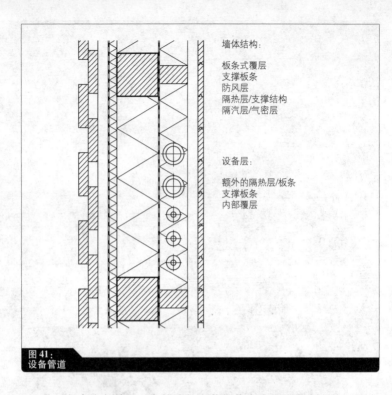

图41:
设备管道

设备层

对于木建筑来说，通常外墙的安装工作会有一个问题。为了避免渗透闭合的表层，依据建筑科学的原则（汽密/气密），所有的管道、供热和供电设备应远离外墙，组装在内墙上。但是，也不可能完全避免外墙不布置设备，在汽密层之前，它们都应该有自己的隔离措施。管道横截面的空间可通过将板条结构尺寸增至 4~6cm 来实现。这种内层也能够当作额外的隔热层来使用。专设浴室和潮湿房间中应专设设备墙和管道。(见图41)

P71

孔洞

防风层、隔热层、隔汽层和其他确保气密性的元件组成的墙体结构，必须与门和窗户这种孔洞合理连接，以避免结构表层功能折减。木材覆层的过渡和设计也应特别注意。

搭接式覆层的剖面在图 42 左侧有清晰的展现，四周布以框架，牢固地连接在窗框上。木板在过梁和窗台上对角放置，使雨水能够顺利地流下。

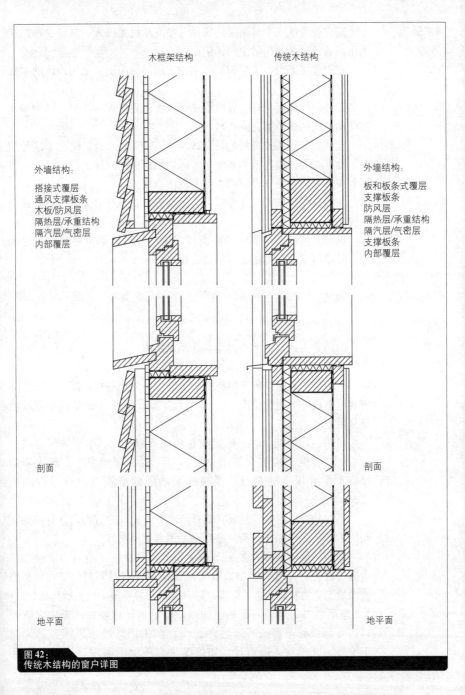

图42：
传统木结构的窗户详图

通风层	窗户处应保证木材覆层的通风。空气从过梁进入，从窗台隔汽层逸出。通风道必须用防虫网保护。
在内部交叉点，窗框和支撑结构之间的缝隙也必须用连续的木板覆盖保护。
隔汽层	和外部防风层一样，隔汽层必须与窗框直接连结。窗户结构和支撑结构之间的缝隙填充隔热材料。
安装板和板条式覆层的传统木墙细部（见图42右）说明了这种情况下没有必要使用连续的外框架，所以窗户被向外移到覆层平面。覆层和窗户之间的垂直缝隙用外部覆层木板覆盖。
防风层	防风膜用板条固定在窗框上。窗框必须企口在过梁上，以连接垂直覆层。
内部，窗户框架包含着深窗框。隔汽层和内部覆层直接与之相连。外窗与承重结构用镀锌钢架连接。
P73	**内墙**
P73
承重，非承重结构	结构
内墙分为两类：承重内墙和非承重内墙。
承重内墙承担自重和顶棚以及屋顶的荷载。加固墙也属于承重墙。它们整体是建筑承重结构的一部分。像外墙一样，用加肋或支架将其建造成一个刚性墙面（见"承重体系"章节）。非承重墙的主要功能是分隔空间。
通常，内墙同外墙一样建造在相同的栅格上，采用同样的结构体系。支撑肋贯穿整个楼层高度，与门槛和顶梁相连。承重墙的顶梁支撑着顶棚的托梁。顶棚托梁之间的空间与内墙结构一样装修。
不像外墙，内墙的主要功能是隔声和防火。建筑物内部分割热空间的墙体不要求隔热，而且厚度也不用考虑隔热需求。
隔声	墙体的隔声效果首先由单位面积的重量决定。内墙的隔声效果由高密度板材厚度提升，如石膏或硬纸板。空腔阻尼，即在面板空腔内填充矿物质或者椰子纤维，隔声效果非常明显。填充墙体的1/2～2/3厚度即可，余留一定的空间给设备管道。内墙在一侧应保证开口，直到设施全部安装以后再封闭。如果用纤维素隔声，需先封上空腔，才能吹入隔声片。如果要求高质量的隔声效果，那么就

要采用双叶墙。这里介绍了单叶墙如何防止声音从墙壁一侧传播到另一侧。

安装固定

内墙与外墙连接处，要特别注意。外墙和内墙不用被斜向锚定，内部覆层的基底必须固定在内部拐角。

在木框架结构中，墙体通常独立地与结构栅格连接，由房间功能决定如何划分空间。外墙加建两根柱子，以确保与外墙的斜向连接，同时为内墙覆层的固定提供途径（见图22、45页和图43）。

传统的木结构中，内墙的分配与建筑栅格的承重柱所对应，所以墙的连接处落在木框架柱的轴线上。柱子两侧用内墙覆层固定的墙体螺栓群加固（见图44）。

在原木结构中，内墙和外墙的连接与外部拐角类似，采用搭接节点。两扇墙由嵌齿或者燕尾榫锚固抗拉（见"原木结构"章节）。内墙的燕尾榫有一个显著特征，末端纹理从外部可见（见图45）。

> 提示：
> 在德国比较潮湿的地域，必须使用绿色的特殊胶粘的木产品面板或者浸渍壁板。石膏纤维板不需要额外的处理就可使用。假如墙肋间距大于42cm，需要两层墙壁面板来支撑瓦片。

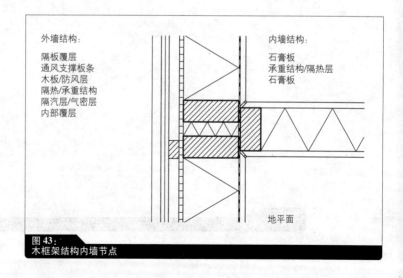

图43：
木框架结构内墙节点

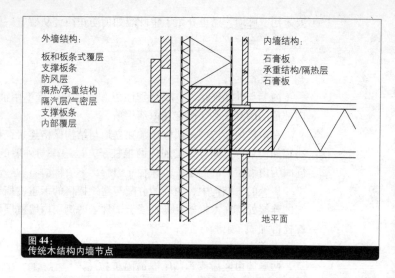

图 44：
传统木结构内墙节点

外墙结构：
板和板条式覆层
支撑板条
防风层
隔热/承重结构
隔汽层/气密层
支撑板条
内部覆层

内墙结构：
石膏板
承重结构/隔热层
石膏板

地平面

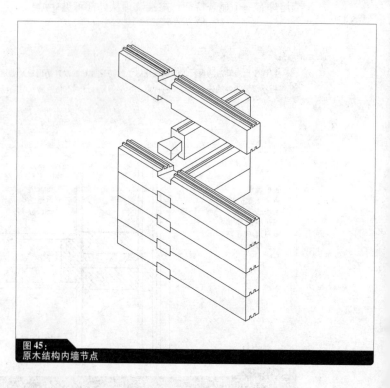

图 45：
原木结构内墙节点

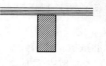 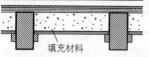 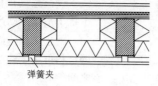

填充材料

弹簧夹

单层

结构：
地板表层/可用表面
顶棚覆层
顶棚梁

双层

结构：
地板表层/可用表面
找平层
碰撞声隔声层
顶棚覆层
填充层
滴水保护膜
顶棚托梁
内部覆层（板条上）

三层

结构：
地板表层/可用表面
找平层
碰撞声隔声层
顶棚覆层
空腔隔热层
地板梁
支撑板条（弹簧）
内部覆层

图46：
单层，双层和三层结构

P76　　　　　　顶棚

木质顶棚可以用托梁或者实木建造。因为用料经济，托梁顶棚在传统的狭板建筑中占据了主导地位。由预制实木组件组装的顶棚近年来变得愈发普遍，如同板式结构一样组装快捷。

P76　　　　　　托梁顶棚

隔声

托梁顶棚结构中隔声是一个重要因素。我们将其区分为结构挤压或<u>碰撞声</u>和<u>空气传播声</u>。地板上的来回走动是产生碰撞声的一种方式。空气传播声来源包括房间里的说话声、收音机发出的声音、电视声或其他类似的声源。

P76　　　　　　结构

根据声音传播的质量可区分单层、双层和三层顶棚建筑。

碰撞声

对于<u>单层</u>结构而言，地板表层和支撑结构中间直接连接，因此当地板上有人走动时，声音会自由地传播。

在<u>双层</u>建筑中，表层和支撑结构之间由碰撞声隔声层隔离。地板表层需要它自身的支撑结构——找平层，其在木结构中是以干燥状态出现的。例如，干燥的找平层可能是由双层石膏纤维板或者木屑板制成。（见图46、图48和"木制品"章节）

空气传播声音	三层结构用于特殊隔声需求，例如将卧室与特别嘈杂的区域分隔。铰接悬挂顶棚用弹簧扣（金属薄片弹性条）与托梁下部相连，干扰了平板振动产生的声波的直接传播。
	阻止声音传播的首要措施是增加单位面积的重量。相对较高密度的材料可用于建造顶棚，或铺设在顶棚覆层上，或填充在顶棚托梁之间。
填充	特殊的干砂就是一种可能的填充材料（见图46）。它在顶棚托梁之间从上部灌进覆层，在很大程度上降低了顶棚托梁的可视高度。为确保顶棚振动时砂子不会滴漏下来，应使用分隔片。
	另一种在顶棚覆层上增加顶棚重量的方法是铺设混凝土铺路石，并将其牢固地粘结在覆层上。然后再铺设余下的地板结构、碰撞声隔声层、找平层和覆层。

P78	托梁
间距	设计实木托梁顶棚时，应注意尽量使梁与房间的短边平行布置。实木托梁的最大间距约为5m，两跨及以上的连续梁比单一托梁更经济。
尺寸	高度比宽度更影响实木托梁的承载力。因此，实木托梁应当竖直安置，来相对经济地利用木材横截面。宽高比通常为1:2或更多。方木的最大高度为240~280mm。
TJI托梁	TJI托梁是非常有效且价格合理的体系，几乎应用于美国所有的木结构中。名字起源于生产商Truss Joint MacMillan Idaho。双T形结构由实木或者胶合网形OSB板粘结而成的饰面胶合板弦组成（见"木产品"章节）。它的质量较轻，但是承载力很强。因为组件非常高，所以管道贯穿托梁相对容易，工厂就是按照这个目的来加工间距的。TJI托梁顶棚通常是下侧覆层（见图47）。

> 提示：
> 如果地板覆层被金属板和侧墙完全分开，我们称其为飘浮结构。需要确保声音不通过墙体传播，边缘也不行。因此，应建造外围条形隔声板，和水平安置的碰撞声隔声层。（见156页的图49）

> 提示：
> 这里提供一个粗略计算木托梁顶棚尺寸的公式：
> 托梁高度h=跨度/20
> 只能用层压板或木产品来获得更高梁高。

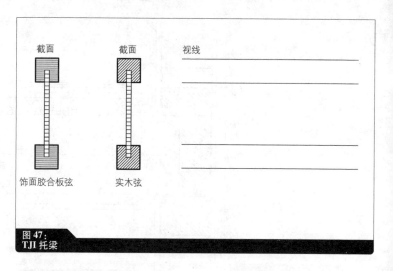

图47:
TJI 托梁

(截面 饰面胶合板弦 / 截面 实木弦 / 视线)

托梁间距	托梁之间一般留有 60~70cm 的间距。但是木结构中,应使节点间距与建筑栅格相吻合,以便顶棚荷载直接传递到承重柱上。
	在大多数的层状木结构体系中,墙体支撑着托梁。必要的承重厚度可以用节点高度×0.7 计算得出。
承接梁	如果承重托梁被设备井、烟囱或者是楼梯贯穿而中断,可使用承接梁。木屋顶桁架的椽子也可使用承接梁(见"屋顶")。承接托梁与其他托梁齐平,通过镶榫连接。节点处由金属夹子固定,防止凸榫被拔出。为了防止火灾,烟囱和木托梁之间需留有 5cm 的净距。

图48:
顶棚体系——托梁顶棚,托梁骨架式结构,TJI 托梁

155

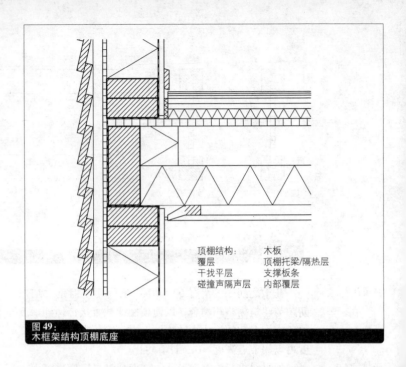

图49:
木框架结构顶棚底座

顶棚结构： 木板
覆层 顶棚托梁/隔热层
干找平层 支撑板条
碰撞声隔声层 内部覆层

底板

图49介绍了标准木结构的详细施工情况，外围托梁形成一种顶棚托梁框架，像整体系统中的外围系梁一样，当担任张压构件时，以防细小托梁滑向一侧。为在静态下达到平板效果，托梁上必须铺设加劲面板，如胶合板。需要注意偏置面板节点，以确保它们能够像砌体结构那样牢固连接。

隔汽层

连续的隔汽层贯穿整个外墙造成了一个特殊的问题。为了防止底座处隔汽膜的断开，隔汽膜应该布置在整个顶棚周边，并连接到下层地板。注意隔汽层在底座处不能中断或者变薄，防止墙体或者底座处形成冷凝水（请看"建造科学"章节）。

托梁吊架/
托梁支撑

图50的细部表明传统木结构施工中托梁在两墙之间悬挂。这种对接节点需要托梁钢吊架或者托梁支撑。建材市场上有各种尺寸吊架或支撑产品，可适用于建筑静力学。

托梁高度的连续围栏使得对接节点的连接成为可能。在这种情况下，隔汽层可在同一平面垂直贯穿，但必须连接到托梁吊架或者托梁支撑上以保证整个结构的汽密性。由于木顶棚托梁可见，细部图显示了托梁支撑被采用。

内部底板　　　如果托梁不能够如图示那样悬挂，而是有意地放置在墙上，可在外墙前沿设置内部底座，形成另一个内部支撑平面，同时也可作为一个额外的隔热层、设备层或者内部覆层的子结构（见148页图41）。

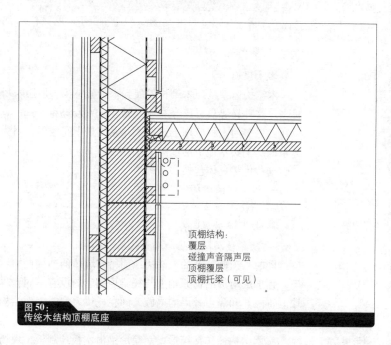

顶棚结构：
覆层
碰撞声音隔声层
顶棚覆层
顶棚托梁（可见）

图50：
传统木结构顶棚底座

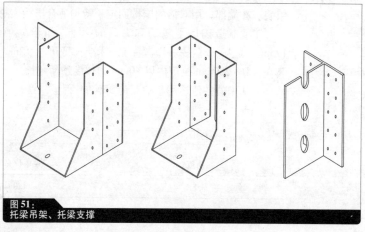

图51：
托梁吊架、托梁支撑

> **提示：**
> 暴露的托梁吊架的使用取决于木托梁顶棚是否在下面设置覆层。否则，就会将托梁支撑用于暴露托梁。这些安装在托梁顶部的狭槽中，通过横向销钉将托梁与外墙固定（见图28和图48中）。

实木顶棚

木结构体系从原木结构发展到木骨架结构，所用木材的数量逐渐减少。这种趋势最近已经有所逆转（见"板式结构"章节）。随着技术的革新，更多的木材应用于建筑中，尤其是中间楼层，因为一系列新的体系证实了实木地板的**优势**。

这些优势具体包括：

— 缩短组装时间；

— 简化工业生产；

— 较薄的横截面；

— 改善楼层间隔热和隔声效果。

同时，顶棚下部的光滑边缘使得与墙体的连接更加容易。

顶部的附加结构原则上与托梁顶棚上的没有不同，尽管不需要采取改善隔热和隔声效果的措施，即在托梁之间或表面结构之间填充材料以增加单位面积重量。

箱形梁

实木结构还包括**箱形梁**。箱形梁由木板预制板，现场像地板一样组装。在端部，用双倍的镶榫固定，特别适合搭接大跨结构。

工业制造确保了精确和高质的标准。箱形梁单元标准宽度为195m（疑为mm。——译者注），标准长度可达12m。根据跨度，梁高在120~280mm 范围内以20mm 的梯度逐渐增长。

表9中的数值适用于荷载为 $3kN/m^2$ 时。

表9：箱形梁尺寸标准值

跨度	单位高度
3.8m	120mm
4.5m	140mm
5.2m	200mm

边缘粘结结构

边缘粘结结构体系利用树干边板和不适用于梁和方木的小尺寸木材。木板纵向放置，直立连接，而不用侧钉按固定的顶棚或墙体构件的钉钉方案来粘结。纵向节点须交错布置。这些带特殊企口的组件固定在整个截面之上（见图52）。

边缘粘结组件需注意防潮，尤其是在施工期，因为木板的渗透水能够造成深度膨胀而不利于安装。

边缘粘结组件的标准尺寸为：厚度 100～220mm，宽度可达 2500mm，长度可达 17m。

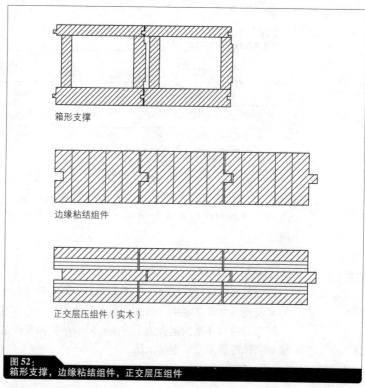

图52：
箱形支撑，边缘粘结组件，正交层压组件

表10：
边缘粘结组件尺寸标准值

跨度	单位高度
3.6m	100mm
4.3m	120mm
5.0m	140mm

正交层压组件

表 10 中的标准值适用于荷载为 $3kN/m^2$ 时。

正交层压组件包含数层带鸠尾榫的软木板，这些木板厚度为 17mm 或 27mm，相互正交粘结。正交可保证形式上的稳固性，因此适用于建造墙体结构。表层也能够用其他木产品来提高表面性能。

根据表层和层数，木板厚度为 51～297mm，最大宽度为 4.8m，最大长度可达 20m，以满足特殊用途。用这种材料，墙体能够建造至 4 层高。

表 11 中的值适用于荷载 $3kN/m^2$ 时。

表 11：正交层压板尺寸标准值

跨度	单位高度
3.8m	115mm
4.6m	142mm
6.4m	189mm

依据类似的<u>实板结构</u>原则，以上三种实木顶棚体系也适用于墙体的实木材料。市场上新产品的数量和新体系越来越多，墙体结构原理与现代原木结构相仿。实木承重墙外侧装有隔热层和防风罩。

屋顶

大多数砌体建筑都包含一些木结构，坡屋顶通常采用木桁架，总的来说是独立式住宅中。但进一步说，砌体结构的木屋顶桁架代表了一种包含两套体系的混合施工方法：轻型结构和实体结构；干式施工法和砌体建筑采用的湿加工法。

> 提示：
> 关于以下术语的解释，读者可再次参考本套基础教材中的《屋顶结构》一书，中国建筑工业出版社 2010 年出版。

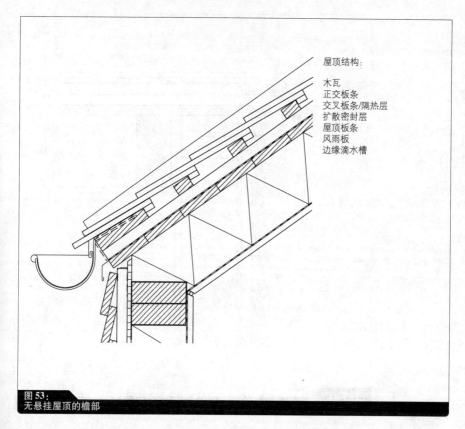

屋顶结构：
木瓦
正交板条
交叉板条/隔热层
扩散密封层
屋顶板条
风雨板
边缘滴水槽

图53：
无悬挂屋顶的檐部

P85　　　　　　坡屋顶

木结构中，坡屋顶和墙体的施工原则统一。

分层　　　　功能和分层顺序，从外到内完全一样：带子结构的<u>防风雨层</u>，<u>密封层</u>或防风层，<u>隔热层</u>，<u>隔汽层</u>和<u>内部覆层</u>。但目的是从屋顶到墙体的过渡点将各层连接起来形成延续的表皮，完成所有科学施工功能，而不中断或削弱。

锚固　　　　此外，形成的<u>屋顶荷载</u>自重和冬天的雪荷载，必须传递到外墙。同时，屋顶应锚固在外墙，抵抗<u>风吸力</u>，而屋顶尤其容易受到风吸力影响。但除了大量的技术要求以外，设计要求也不能被忽略。屋檐和建筑边缘对建筑风格有特殊的影响，主要问题是屋顶是否应该悬挂。

施工科学　　图53的剖面细部展示了屋顶到墙体过渡相对简单的方法，各层非常容易连接成整体。两个剖面的内墙板覆层在过渡段的边缘被修剪平齐。如果房间内墙体颜色和节点一致的话，胶粘密封层很难被察觉。

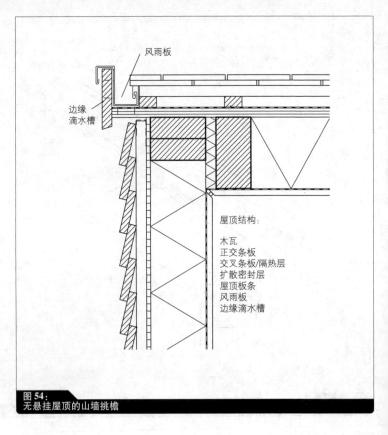

图54：
无悬挂屋顶的山墙挑檐

将护墙板粘结到屋檐和山墙挑檐内部拐角，能够完全闭合整个屋顶空间的隔汽层和气密层。

承重墙柱和屋顶椽子之间的<u>隔热层</u>直接相连，纵墙和两端山墙顶梁形成了上部的围合。

同砌体结构一样，檐部不需要专门设置檩条，因为<u>椽子</u>能够直接搁在<u>木框架墙</u>上。端部山墙和外围椽子之间缝隙填充隔热材料的边缘应特别注意，避免产生严重热桥。

底座
,🛈

🛈
提示：
山墙不像其他墙体那样，它的顶梁不是水平放置，而是沿屋顶坡度对角安放。平面边缘被正交切割以避免扭曲的横截面（见图54）。

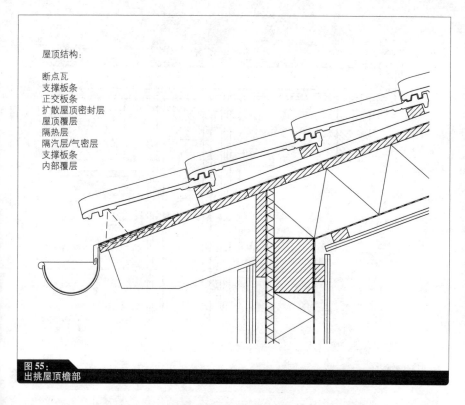

屋顶结构：

断点瓦
支撑板条
正交板条
扩散屋顶密封层
屋顶覆层
隔热层
隔汽层/气密层
支撑板条
内部覆层

图 55：
出挑屋顶檐部

外层	对于屋顶，整个椽子隔热层覆盖着木板和扩散性的屋顶薄钢板（$S_d > 2m$）。墙上的加固胶合板形成围合，保护隔热层。
	屋顶和墙体外层从后侧通风。每个通风道各自独立，有单独的进风口和出风口。
屋顶边缘	没有出挑屋顶强调了建筑物的体量，使之显得更为宏伟。墙体和屋顶同样影响着建筑的整体形象，这种印象由墙体搭接覆层和屋顶平尾瓦之间的比例关系所改观。
	屋顶山墙的边缘容易遭受风荷载，通常用风雨板来应对。由于屋顶覆层禁止端部纹理使用螺栓，所以用镀锌扁钢条将风雨板固定在屋顶覆层上。屋顶表层和风雨板之间的降水由金属<u>边缘滴水槽</u>收集，直接流到檐部，进入屋顶滴水沟。
屋顶出挑	屋顶出挑使建筑物的空间感很不一样。屋顶出挑与建筑物主体更加相互脱离，好像有了自己的生命。屋顶表层和墙体覆层的不同材料语言强调了两者各自的风格。

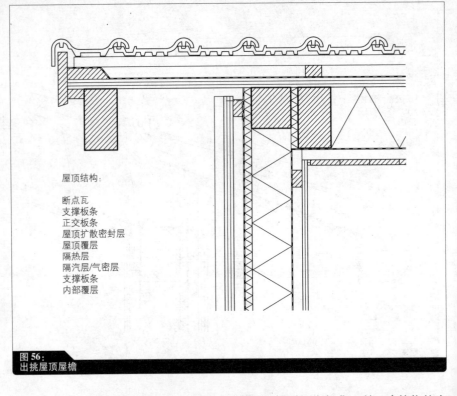

屋顶结构：

断点瓦
支撑板条
正交板条
屋顶扩散密封层
屋顶覆层
隔热层
隔汽层/气密层
支撑板条
内部覆层

图56：
出挑屋顶屋檐

    四周悬挑的屋顶可防止墙体覆层遭受雨水侵袭，利于建筑物的<u>木结构防护</u>，相应的缺陷是椽子贯穿了外表。但斜面能够保证椽子端部被屋顶表层很好地保护，而且较细密的木材端纹不会直接暴露于风雨中。

墙节点   为了避免对应垂直的板和板条覆层与椽子的繁复劳动，覆层在较低端断结。椽子之间插入木板（见图55和图56），同时固定了覆层上端。外部覆层后的通风道由内板和覆板之间的空间排风。

边缘   边缘处，墙体覆层沿屋顶边缘铺设，延伸到屋顶覆层下，需留有2~3cm的缝隙来确保通风道的空气顺利逸出。

    只有突出的檩条能够造成山墙边的屋顶挑檐。檩条支撑着外部的椽子即<u>檐口椽</u>，檐口椽不能由建筑物内部支撑，而是由纵墙的木框架支撑。超出山墙的部分为檐口槽，继续支撑着檐口椽。按照突出宽度来确定木框架的尺寸，将方的横截面用立梁形式来代替更好。

    图56的檐口细部与平屋面板瓦不同，因为不需要设置<u>边缘滴水</u>

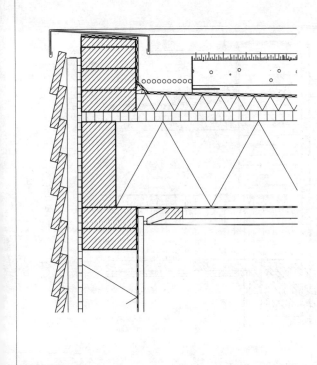

屋顶结构：

贴附层/过滤层/排水层
根基保护层
屋顶密封层
倾斜保温层
隔汽层
屋面板
隔热层/梁
隔汽层
支撑板条
内部覆层

图57：
平屋顶边缘托梁顶棚——木结构施工

槽，而是通过一种特殊形状的檐瓦排水。檐瓦和风雨板围合并保护着屋顶覆层。风雨板边缘与覆层上下边缘的加固条用螺栓连接固定。

P91　　　　　平屋顶

平屋顶是现代建筑的一个重要特征。混凝土结构索来广泛使用平屋顶，如今现代木结构施工也开始采用。

女儿墙　　　图57的细部展示了外部表皮延伸至屋顶边缘，成为斜面覆层，在平屋顶的情况下有时也叫女儿墙。因为有屋顶女儿墙挑出屋顶至少10cm以外，所以不需要像坡屋顶山墙一样用风雨板来保护屋顶。女儿墙和敞顶式立面覆层由向内倾斜的金属薄片覆层保护，立面通风道由此排气。

底座　　　　屋顶顶棚的施工和固定方法与中间屋顶棚一样。(见图49）屋顶女儿墙由两层水平正交的木材组成，使人联想起中间层连续的门槛。

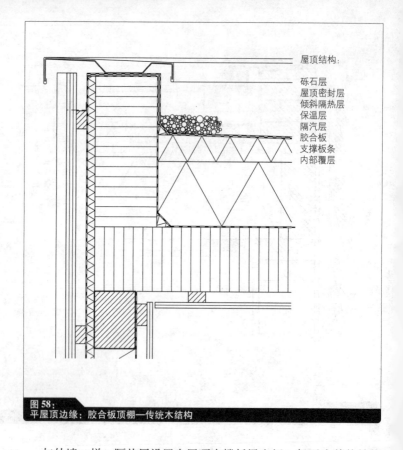

图58：
平屋顶边缘：胶合板顶棚—传统木结构

与外墙一样，隔热层设置在屋顶支撑托梁之间。但因为其他结构设计成不通风平屋顶，屋顶隔热层内的隔汽层作用十分关键。屋顶覆层上还额外安装了一层斜置的隔热材料，来确保从固定的屋顶密封层一直到屋顶集水沟的连续坡度。这通常可以用堆积的粒状材料来完成。这层隔热层其后也将下面的屋顶隔热层中可能的弱点连接起来。

轻型平屋顶储存空间较少，可以做成绿色屋顶。出于防火考虑，不窄于50cm的砾石条层应将植物与平屋顶边缘和木结构组件分隔开来。

图57和图58的平屋顶边缘只各显示出屋顶边缘外部设计和屋顶结构本身诸多详细方法中的一种。

还应当记住，虽然不通风结构是一个非常普遍的解决方案，但是平屋顶可以像坡屋顶一样通风建造。

用边缘胶合板建造的屋顶顶棚（见"顶棚"一章）结构和细板非常接近钢筋混凝土平屋顶。

实木建筑有利于加强保温效果和提高屋顶的存储率。边缘胶合组件由木墙框架支承；到屋顶女儿墙的斜坡也是由胶合构件建成的。

结构

实木顶棚上应使用通风屋顶结构，类似于钢筋混凝土平屋顶。这里，隔汽层铺放在屋顶顶棚之上，保温层之下。墙体和屋顶的隔热层只能通过实木构件间接连接。承重结构外的外部保温层可防止热桥的形成。

结语

在结束本卷之时，须再次说明此书是突出木材的建筑材料特性，在戈特弗雷德·森佩尔著作《Style》的构造章节中，他称之为所有木结构建筑的原始材料。

木结构施工意味着严格按照建筑的逻辑和清晰易懂的法则建造。不同于实体结构，木结构中力的传递可以直接得知和研究。

对木结构施工的理解，有利于扩展理解建筑师常用的其他结构体系。钢结构和木结构的钢筋节点和平面就很相似。金属和玻璃立面也有类似的情形，取代了木结构施工中的柱和栏杆组合方式。甚至混凝土作为浇筑材料，也采用了木结构的构造原理，承重系统由柱、梁和钢筋组成。假如把承重钢筋的加固功能想像为木梁，那么钢筋混凝土顶棚的承重效果就更容易被理解。

所以，许多建筑学课程都是以木结构施工开篇。它有利于理解建筑基本原理，而且诸多特殊领域都需要用到它的知识。

附录

标准

普通木结构

DIN EN 338	木结构—强度等级
DIN EN 384	结构木材—力学特性和密度的测定值
DIN V ENV 1995-1-1 Eurocode 5：木结构设计；通则和建筑规程中的 1-1 部分。	
AS 1684.1-3 1999	木住宅—框架结构—设计标准
AS 1720.1997	木结构—设计方法

建筑木材

DIN EN 338	承重结构木材—强度等级
DIN EN 384	承重结构木材—力学特性定义、刚度和容积密度的测定值
DIN EN 1912	承重结构木材—强度等级—视觉分类和木材类型分级

木结构防护

DIN EN 335	木材和其衍生材料的耐久性；生物腐蚀危害等级定义
DIN EN 350	木材和木产品的耐久性
DIN EN 351	木材和木产品的耐久性—防腐处理实木

美国标准

建筑通用规范，UBC
建筑通用规范，第 25 章 木结构
建筑通用规范，第 5 部分第 25 章 木结构手册—插图说明
木结构—住宅框架施工，美国农业部林业司
木结构手册，美国农业部林业司

参考文献

American Institute of Timber Construction (AITC): *Timber Construction Manual*, John Wiley & Sons, 2004

Werner Blaser: *Holz-Haus. Maisons de bois. Wood Houses*, Wepf, Basel 1980

Francis D.K. Ching: *Building Construction illustrated*, 3rd edition, John Wiley & Sons, 2004

Andrea Deplazes (ed.): *Constructing Architecture*, Birkhäuser, Basel 2005

Keith F. Faherty, Thomas G. Williamson: *Wood Engineering and Construction Handbook*, McGraw-Hill Professional, 1998

Manfred Hegger, Volker Auch-Schwelk, Matthias Fuchs, Thorsten Rosenkranz: *Construction Materials Manual*, Birkhäuser, Basel 2006

Thomas Herzog, Michael Volz, Julius Natterer, Wolfgang Winter, Roland Schweizer: *Timber Construction Manual*, Birkhäuser, Basel 2003

Theodor Hugues, Ludwig Steiger, Johann Weber: *Timber Construction*, Birkhäuser, Basel 2004

Wolfgang Ruske: *Timber Construction for Trade, Industry, Administration*, Birkhäuser, Basel 2004

William P. Spence: *Residential Framing*, Sterling Publishing Co., New York 1993

Anton Steurer: *Developments in Timber Engineering*, Birkhäuser, Basel 2006

图片出处

图 11, 15, 18, 24	Theodor Hugues
图 14	Ludwig Steiger
图 23	Jörg Weber
图 36	Anja Riedl
图 39	Anja Riedl/Jörg Rehm
图 48	Ludwig Steiger/Jörg Weber
图页 12	Johann Weber
图页 28, 67	Jörg Weber
图页 87	Architekturbüro Fischer + Steiger
所有线图	Florian Müller